AI & You:

WILL HUMANITY SURVIVE THE FUTURE WE'RE BUILDING

by, Faisal I. Siddiqi

AI & You:
WILL HUMANITY SURVIVE
THE FUTURE WE'RE BUILDING?

Copyright © 2024
All rights reserved

First Edition – October 2024

Author: Faysal Siddiqi
Published by: SizLink Communications
                1.855.749.5465

No part of this book may be used or reproduced in any manner whatsoever without the prior written permission of the publisher, except in the case of brief quotation embodied in reviews.

Library and Archives Cataloguing in
Publication information is available.

ISBN: 978-1-0691137-6-4

This book is dedicated:

To the younger generations who stand on the brink of exploring the world of artificial intelligence. May you navigate this evolving landscape with curiosity, creativity, and critical thinking.

To the future generations who will master these technologies, I hope you wield them wisely and responsibly, shaping a world where humanity thrives alongside the machines you will create.

"

Machine learning and AI are going to be so important. Every single large company that isn't using AI in some way will be disrupted.
— **Jeff Bezos**

AI, in the long run, is the biggest threat to humanity. But in the near term, it's going to make a lot of difference in terms of helping to make the world more efficient.
— **Bill Gates**

With artificial intelligence, we are summoning the demon.
— **Elon Musk**

Artificial intelligence is not a substitute for human intelligence; it is a tool to amplify human creativity and ingenuity.
— **Fei-Fei Li**

Contents

Preface: Why Am I Writing This Book? 1

Chapter 1: What Really is AI? 7

Chapter 2: How AI Works: The Magic Behind the Curtain 14

Chapter 3: The Rise of AI in Everyday Life 18

Chapter 4: Myths and Misconceptions About AI 28

Chapter 5: AI's Learning Curve 34

Chapter 6: AI in the Workplace – Reshaping Jobs and Industries 40

Chapter 7: AI and the Global Economy– Redefining Growth and Power 48

Chapter 8: AI and Society–Culture, Education, and Ethical Dilemmas 55

Chapter 9: AI and the Workforce – Disruption or Transformation? ... 62

Chapter 10: AI and Governance – Regulating the Future 71

Chapter 11: AI and Privacy – Navigating the New Data Landscape ... 78

Chapter 12: AI and Human Rights 87

Chapter 13: Coexisting with Machines Finding Balance in an AI-Driven World 92

Chapter 14: The Future of Human Identity in an AI-Driven World 99

Chapter 15: AI for Good: Potential and Possibility 104

Chapter 16: Can We Survive the AI Revolution? 109

Chapter 17: Your Role in the AI Future 114

Epilogue: The Next Chapter for Humanity 119

Preface: Why Am I Writing This Book?

I've been wanting to write this book for long time, ever since the usage of AI has exploded, becoming an integral part of our daily lives in ways that are both subtle and profound. When I see how AI has moved into our daily lives, guiding us on our routines, creating probable choices by listening to our conversations, and influencing everything from music recommendations to shopping, I can't help but reflect on its pervasive impact. Kids are using AI in their classrooms and projects, and the corporate world is adapting, with employees leveraging these tools to improve productivity. However, this rapid integration raises critical questions: What will happen to humans as we become increasingly dependent on AI shortcuts? What effects will this reliance have on our brains when we stop stimulating our natural curiosity to seek answers, solve problems, and innovate? Another common concern looms large: What will happen to future jobs in an AI-dominated landscape? How will the evolution of human industry and the world unfold?

In the corporate world, AI is helping employees improve productivity by automating routine tasks, analyzing data, and offering insights faster than any human could. But it was during a conversation that the real spark for this book came to life. I was talking with a group of doctors and teachers, and they casually brushed off the idea that AI could ever be useful in their workplaces. "AI can't do what we do," one of them said, confidently dismissing it. That made me pause. I realized that even though AI is rapidly becoming a game-changer, there's still so much skepticism and misunderstanding, even in fields where it could have a profound impact.

As I observe the younger generation today, it's striking how extensively they engage with AI in their academic pursuits. From elementary school to university, students are increasingly relying on AI tools to assist with assignments and projects. Whether it's using AI-driven writing assistants to enhance their essays, employing language translation apps to aid in learning new languages, or leveraging data analysis tools for research, these technologies have become integral to their educational experiences. This reliance not only streamlines their workflow but also enriches their learning by providing instant access to information and resources that would have been unimaginable just a few years ago.

Similarly, in the corporate world, employees across various industries are embracing AI to optimize their work processes. Managers utilize AI analytics to make data-informed decisions, predict market trends, and enhance operational efficiencies. In fields such as finance, healthcare, and marketing, AI systems analyze vast amounts of data, helping professionals uncover insights that drive strategic initiatives and improve customer engagement. The automation of routine tasks allows workers to focus on more complex and creative aspects of their roles, ultimately enhancing productivity and innovation. This trend reflects a significant cultural shift: technology is no longer just a tool but a collaborative partner in the workplace and the classroom.

Reflecting on my own experiences, I think back to my time at Rogers over a decade ago, where I was part of the Customer Base Management (CBM) team. We were building predictive models for customer churn and analyzing user behavior—a formative experience in the early stages of AI and data modeling. This early foray into data modeling was a significant moment in the evolution of AI and highlighted its potential to transform how businesses understand and engage with their customers. The insights gained from this experience not only underscored the power of data-driven decision-making but also brought to light the ethical considerations of using such technology.

But this transformation raises many questions that we can no longer afford to ignore. What will happen to humans as we become more dependent on AI for even the simplest of tasks? What happens to our creativity and problem-solving abilities when we let machines do the thinking for us? Are we unknowingly allowing our brains to stagnate, relying on AI shortcuts instead of seeking answers through curiosity and research? The rapid adaptation of AI in the workplace also brings with it a common worry: what will happen to jobs, industries, and livelihoods as machines increasingly take over tasks that were once the domain of humans?

Already, AI is transforming industries at a rapid pace. From automating legal agreements and consultations to handling tasks in media, such as generating news articles or editing video content, this new tool is reshaping entire professions. Many jobs that require repetitive processes, whether in finance, law, or journalism, are now within the capabilities of AI systems. What was once done by humans—drafting contracts, conducting consultations, or even providing customer service—is now being done by algorithms in minutes.

This raises a fundamental question: What does the future hold for human employment as AI continues to evolve? How will industries adapt to these changes, and how will people find their place in a world where machines increasingly take over tasks that require not just labor but intelligence?

Writing this book is my attempt to make sense of these questions. The idea of AI replacing human effort is both exciting and frightening. I want to explore not only how AI works and what it can do but also how it is reshaping our world—and what this means for our future as human beings. What will become of us when machines do so much of the heavy lifting? Will AI elevate humanity to new heights, or will we lose something essential in the process?

This book aims to explore these questions in depth, breaking down what AI truly is, how it's being adopted, and separating the myths from the reality. Because as much as AI offers promise and opportunity, we also need to be aware of the challenges it brings—and more importantly, how we as humans can thrive alongside it.

As we embark on this exploration of AI, I invite you to join me in examining not only its current applications but also the profound implications it has for our future. Together, we can navigate this evolving landscape and advocate for a future where technology enhances our lives while remaining mindful of our humanity.

Part One:

Understanding AI

Chapter 1: What Really is AI?

AI or **Artificial Intelligence** - The ultimate buzzword. You've probably heard it everywhere by now. Whether it's your tech-savvy friend going on about how AI is the future, or a news article predicting that robots will steal all our jobs—AI is a term that's thrown around like confetti at a tech party. But if you're honest, you might still be thinking: "Okay, so what is it really?" Is it some sort of robot overlord waiting to take over the world? Or is it just an algorithm running in the background of my Netflix account, figuring out why I watched an entire season of that reality show I swore I hated?

The truth is, AI isn't one thing; it's many things, all bundled together under one shiny, techy label. It's not some magical brain that's going to outsmart humans tomorrow (well, not yet), but it is incredibly powerful and already shaping our world in ways you might not realize. Let's break it down.

Understanding AI: A Simple Definition

Let's break it down in the simplest way possible. AI is a branch of computer science focused on building machines that can "think." But when we say "think," it doesn't mean the way humans do. AI doesn't have emotions, consciousness, or intentions—it's simply using data and rules to make predictions, solve problems, or automate processes.

A common way to understand AI is to think of it as a set of tools. Imagine you're trying to fix something in your home. You have a hammer, a screwdriver, and a wrench. AI, in a sense, is like those tools. It's designed to solve specific problems, and depending on the problem, you use the appropriate "tool" or technique.

AI: Beyond the Hype and Sci-Fi

Let's start by clearing up some misconceptions. When you hear AI, you might picture a robot that talks like a human, looks like a human (or like your vacuum cleaner on steroids), and maybe even thinks like a human. Blame Hollywood for that image—movies like Terminator and Ex Machina love to show AI as sentient beings ready to rebel or solve existential problems. The truth is a little less glamorous. Real AI is more like an extremely efficient problem-solver, designed to make decisions, recognize patterns, and handle large amounts of data. It's more Watson the IBM supercomputer than R2-D2.

Think of AI like a really smart assistant—but one that's laser-focused on doing a specific job, not an all-knowing genius who can do anything. For example, that spellcheck in your Word document? AI. The suggested playlist on your Spotify? AI. When your phone somehow knows you're craving pizza and gives you a 20% discount on Domino's in an ad? That's AI too. It's in the little things that save us time or make our lives easier, even if we don't notice it.

Automation vs. AI: What's the Difference?

Okay, so AI is smart, but here's a key point—AI isn't just fancy automation. What's the difference? Automation is like that coffee machine in your office that makes your latte the same way every time without fail (unless it breaks). It's programmed to do one thing, and it does it well. But AI is like a coffee machine that can look at your tired face, figure out what kind of day you're having, and automatically give you a double shot espresso because it "learned" you needed a boost. It's smart, flexible, and can adapt.

To put it another way: automation is about pre-programmed tasks, AI is about learning and adapting. AI can make decisions on the fly based on data. It's like that friend who learns over time that you only pretend to like pineapple on pizza and then quietly stops offering it at dinner parties. AI gets smarter the more it's used, picking up on patterns and making predictions—often eerily accurate ones.

AI's Roots: From Dream to Reality

You may think AI is a product of our internet-driven, futuristic world, but the dream of AI has been around for much longer. In fact, the term **"Artificial Intelligence"** was first coined in 1956 at a conference held at Dartmouth College. Back then, AI researchers weren't building fancy smartphones or virtual assistants; they were trying to get machines to solve logical puzzles or play chess, and it was a pretty slow process.

Imagine the earliest AI systems like a toddler learning to walk. Sure, they could do something impressive like add two numbers or follow basic instructions, but they were far from running marathons. AI back then was rule-based, meaning someone had to program every single step. Want the computer to recognize a cat picture? Well, you'd better input a lot of cat pictures and program every possible variation of "cat." Not exactly convenient.

It wasn't until the late 1980s and 1990s that AI got a major boost with something called machine learning. This was the moment AI got its first real sense of independence. Now, instead of needing to be told every single step, AI could start figuring out patterns on its own—like realizing that not every cat looks the same but they all have some similar features (furry, whiskers, judgmental eyes).

Narrow AI vs. General AI: The Battle of the Brains

Right now, AI exists mostly in what's called **Narrow AI.** This means it's great at one thing and one thing only. Think of it like that colleague who's a total Excel whiz but has no idea how to use PowerPoint. AI today is specialized, whether it's diagnosing medical conditions, recommending movies, or driving a car. It's highly efficient at its task, but if you asked it to step outside its comfort zone, it would crumble.

The holy grail, however, is **General AI**—an AI that's not only a movie-picking genius but also your personal doctor, lawyer, chef, and maybe even life coach. This is the dream (or the nightmare, depending on your view), where AI isn't limited by one task but can perform any intellectual job a human can. But don't panic! We're not there yet. General AI is still the stuff of science fiction. For now, your AI can help you order a pizza but isn't going to replace your therapist.

Machine Learning: The Secret Sauce

Now that we've demystified what AI is, let's take a closer look at how it works. The magic behind modern AI is something called **machine learning.** Think of it as teaching a computer to spot patterns and make decisions based on those patterns, just like humans do, but a lot faster and without coffee breaks.

Here's a simple example: If you wanted an AI to identify pictures of dogs, you'd feed it thousands of images of dogs—poodles, bulldogs, dachshunds, and yes, even those bizarre Instagram "dog" filters. The AI starts to learn that all of these images share something in common, like floppy ears or wagging tails, and from that point on, it can identify a dog in a new image it has never seen before. It's not because the AI understands what a dog is—it's just really good at recognizing patterns.

There are different types of machine learning, but the key point is that machine learning allows AI to improve itself without someone programming every possible outcome. AI learns from experience, which is why it sometimes feels like your tech knows you better than you know yourself.

AI in Action: Already in Your Pocket

You don't need to wait for the future to experience AI—chances are, it's already in your pocket, your home, or even your workplace. Here's how AI is playing a role in your daily life:

Your Phone: From voice assistants like Siri and Google Assistant to facial recognition features, AI is what powers the "smart" in smartphones.

Streaming Services: Ever wonder how Netflix knows exactly what you want to binge-watch next? AI analyzes your viewing habits and makes recommendations based on them.

Social Media: Instagram and Facebook algorithms track your interactions to show you the posts and ads they think you'll engage with the most.

Shopping: AI helps online retailers like Amazon show you products you're more likely to buy by analyzing your browsing and shopping behavior.

Healthcare: From diagnostics to personalized treatment plans, AI is revolutionizing medicine, helping doctors make more accurate decisions faster.

You might not have noticed it, but **AI is already embedded in nearly every aspect of your digital life.** It's convenient, yes, but it also brings up questions about privacy, control, and where we're headed as a society.

Chapter 1: What Really is AI?

AI may sound like a distant, futuristic concept, but it's very much a part of today's world. And now that we've stripped it down to its basics, the next chapter will dive into how AI is rapidly being adopted across industries, reshaping jobs, businesses, and life as we know it. But for now, you can take a deep breath—your AI-powered coffee machine isn't plotting world domination (yet).

Chapter 2: How AI Works: The Magic Behind the Curtain

Imagine a newborn baby. It comes into the world with a blank slate, ready to learn and grow. As it interacts with the world, it observes, experiments, and gradually develops intelligence. In a similar way, AI systems learn and evolve through exposure to data.

The Role of Data and Algorithms

Data is the lifeblood of AI. It's the raw material that AI systems feed on to learn and make intelligent decisions. Just as a human brain needs information to learn, AI models require vast amounts of data to train on. At the heart of AI lies a powerful duo: data and algorithms. Data serves as the fuel, powering the engine of AI systems. It's the raw material from which AI models learn and make intelligent decisions.

Data Quality: The quality of data is paramount. Clean, accurate, and representative data is essential for training effective AI models. Poor quality data can lead to biased and inaccurate results.

Data Quantity: More data often leads to better performance, especially for complex tasks like image and speech recognition. However, it's important to strike a balance between quantity and quality.

Algorithms, on the other hand, are the rules and procedures that guide AI systems. They are the brains behind the operation, processing data and making decisions.

Once we have the data, we need algorithms to process it. Algorithms are sets of instructions that tell the computer how to solve a problem. In the context of AI, algorithms are used to extract patterns from data, make predictions, and generate insights.

Training AI Models: A Learning Process

AI models learn through a process called training. To train an AI model, we expose it to vast amounts of data. The model learns to identify patterns, make predictions, and generate insights. There are three primary learning paradigms:

Supervised Learning: In supervised learning, the model is trained on labeled data. This means that the data is paired with correct outputs, allowing the model to learn the mapping between inputs and outputs. For example, to train an image classification model, we would provide it with images labeled with their corresponding categories (e.g., "cat," "dog," "car").

Unsupervised Learning: Unlike supervised learning, unsupervised learning involves training models on unlabeled data. The model's task is to find patterns and structures within the data itself. This is useful for tasks like clustering, anomaly detection, and dimensionality reduction.

Reinforcement Learning: Reinforcement learning is inspired by how animals learn through trial and error. In this approach, an AI agent interacts with an environment and receives rewards or penalties based on its actions. The agent learns to make decisions that maximize rewards and minimize penalties.

The Black Box Problem: Understanding AI's Decisions

One of the challenges with AI is its opacity. As AI models become increasingly complex, it can be difficult to understand how they arrive at their decisions. This is often referred to as the "black box problem."

While AI models can make accurate predictions and classifications, they may not be able to explain their reasoning. This lack of transparency can be problematic, especially in high-stakes applications like medical diagnosis or autonomous vehicles.

To address the black box problem, researchers are developing techniques to make AI models more interpretable. These techniques involve:

Visualizing the internal workings of the model: By visualizing the activation patterns of neurons, we can gain insights into the model's decision-making process.

Identifying the most influential features: By analyzing the importance of different input features, we can understand which factors contribute most to the model's predictions.

Generating human-readable explanations: By using natural language processing techniques, we can generate explanations that describe the model's reasoning in plain language.

The Future of AI: A Collaborative Partnership

As AI continues to advance, it's crucial to foster a collaborative partnership between humans and machines. AI can augment human capabilities, enabling us to solve complex problems and achieve new heights. However, it's essential to use AI responsibly and ethically, ensuring that it benefits society as a whole.

By understanding the fundamental principles of AI, we can harness its power to create a better future for all.

Chapter 3: The Rise of AI in Everyday Life

AI has quietly woven itself into the fabric of our daily lives, often without us noticing. From the moment we wake up and reach for our smartphones, AI is at work, making decisions for us and streamlining our routines. This chapter explores how AI has infiltrated our personal lives, workplaces, and industries, transforming the way we live and work in ways both subtle and profound.

AI in Our Personal Lives: The Invisible Assistant

When was the last time you actually stopped to think about how often you interact with AI in your daily routine? From virtual assistants like Siri and Alexa to Netflix recommendations and spam filters in your email, AI is working behind the scenes to make our lives more efficient and tailored. It's easy to overlook just how much this invisible assistant does for us, but its influence is undeniable.

Let's start with the most obvious example: your smartphone. Every time you unlock it with your face, that's AI. Every time you ask Google Assistant to remind you about a meeting or to play a specific song, that's AI. And every time you see an ad pop up for something you were just talking about—yep, that's AI, too, gathering data, learning your preferences, and making calculated predictions to cater to your interests (or at least, what it thinks are your interests).

Then there's AI in the content we consume. Streaming services like Netflix and Spotify use AI algorithms to track what you've watched or listened to and recommend content that fits your tastes. The result? You get a curated experience that feels personal, even though it's driven by data. AI has become so effective that it often knows what we want to watch or listen to before we even do.

The Corporate World: AI as the New Workforce

The corporate world has embraced AI like never before, and for good reason. It's making businesses more productive, efficient, and even more profitable. AI is being used to automate repetitive tasks, predict market trends, and enhance customer service. What's surprising, though, is just how many industries are benefiting from it.

Take the legal field, for example. In the past, lawyers had to sift through mountains of paperwork to find relevant information for a case. Now, AI can scan thousands of documents in minutes, finding patterns and relevant sections faster than any human could. Some AI tools are even drafting basic contracts and legal agreements, which lawyers only need to review and fine-tune. This speeds up the process and reduces the cost of legal services, though it also raises questions about job security in such fields.

In finance, AI is reshaping everything from risk assessment to stock trading. Robo-advisors, which are AI-driven tools, provide investment advice and manage portfolios based on market data and algorithms. They offer services at a fraction of the cost of human advisors, and because they can analyze vast amounts of data, they often outperform their human counterparts in certain areas.

Customer service is another domain that AI is revolutionizing. Chatbots are now the frontline workers for many companies, handling basic customer inquiries with ease and efficiency. These bots use natural language processing (NLP) to understand customer requests and respond in a conversational way, reducing the need for human involvement in routine matters.

AI in Healthcare: A Doctor's Digital Assistant

One of the most promising areas of AI adoption is in healthcare, where it's helping doctors diagnose diseases faster, manage treatment plans, and even predict potential health risks. AI systems are being trained to analyze medical images, detect patterns in patient data, and identify symptoms that might be overlooked by human eyes.

For example, AI algorithms can now analyze X-rays, MRIs, and CT scans with astonishing accuracy, sometimes even spotting issues that doctors might miss. In oncology, AI is being used to predict how a patient will respond to treatment based on their genetic makeup, allowing for more personalized and effective care.

Doctors are also using AI to assist with surgery. Robotic surgery systems, guided by AI, help with precision tasks in the operating room. These systems can perform complex procedures with incredible accuracy, reducing the risk of human error and improving patient outcomes.

However, just as in other fields, the rise of AI in healthcare raises concerns. Will AI replace doctors? Probably not, but it will likely change how they work, shifting some responsibilities to machines while allowing doctors to focus on more complex and human-centered tasks.

AI in Education: Teaching the Teachers

Education is another area where AI is making its mark, though perhaps more quietly than in other sectors. AI is being used to tailor learning experiences for students, offering personalized tutoring and adjusting to each student's pace and learning style. It's also helping teachers by automating administrative tasks, like grading, and providing insights into student performance that can be used to enhance instruction.

For instance, AI-powered learning platforms can analyze a student's performance in real time and adjust the content to match their skill level. If a student is struggling with a particular concept, the system can offer additional resources, explanations, or exercises until they grasp the idea. On the flip side, if a student is excelling, the AI can fast-track them to more advanced material, preventing boredom and keeping them challenged.

But as with any technology, there are concerns. Some educators worry that relying too heavily on AI could depersonalize the learning experience, and that it might undermine the role of teachers. Others fear that students could become overly dependent on AI tools, losing the ability to think critically and solve problems on their own.

The Job Market: AI as a Job Creator or Destroyer?

Perhaps the most pressing question about AI is how it will impact jobs. For all the ways AI improves efficiency, it also threatens to replace human workers in certain industries. Jobs that involve repetitive tasks—whether it's data entry, assembly line work, or even customer service—are at the highest risk of being automated. As AI continues to improve, the list of vulnerable jobs only grows.

However, AI is also creating new job opportunities. As more companies adopt AI, there's a growing need for people who can develop, manage, and maintain these systems. Roles like AI ethics specialists, machine learning engineers, and data scientists are in high demand. Moreover, AI has the potential to create entirely new industries, just as the internet did decades ago.

The key challenge for the workforce will be adaptation. People will need to learn new skills and move into roles that complement AI rather than compete with it. Industries will have to rethink how they train employees, and governments may need to consider policies to support workers transitioning into new careers.

AI and Entertainment: The Algorithm Behind Your Choices

AI's reach goes beyond the practical applications of healthcare, education, and the workplace—it's deeply embedded in our entertainment. Ever wondered how Netflix seems to know exactly what you'd like to watch next? Or how Spotify curates a playlist that feels tailored just for you? This is the magic of AI at work, shaping what we consume for leisure in ways we often don't even notice.

Streaming services, like Netflix and Hulu, use AI to analyze your viewing habits—what genres you prefer, how long you spend watching a show, and even the time of day you're most likely to binge. With that data, they create recommendations that keep you glued to the screen. In fact, up to 80% of what you watch on Netflix is influenced by its recommendation algorithm.

Music platforms work the same way. Spotify's "Discover Weekly" playlist feels almost eerily accurate because its AI is constantly learning from your listening habits and cross-referencing them with others who have similar tastes. The AI uses vast data sets to predict what new songs you'll enjoy, ensuring you keep coming back for more.

And it's not just passive consumption. AI is also getting involved in creating content. In music, for instance, AI tools like Amper and AIVA can compose original tracks, generating melodies and beats that sound like they were produced by human musicians. Some artists use AI as a collaborative tool, blending their creativity with machine-generated patterns to craft new sounds. While we're still in the early stages of AI-driven creativity, the possibilities are vast—and a bit unsettling for those who wonder if AI will one day replace human artists.

AI and Smart Cities: The Future of Urban Living

Another area where AI is starting to make a significant impact is in the development of "smart cities." These are urban environments where AI and the Internet of Things (IoT) combine to create more efficient, sustainable, and livable cities.

Imagine a city where traffic lights adjust dynamically based on real-time traffic data, minimizing congestion and reducing commute times. Or a city where energy consumption is monitored and optimized by AI, reducing waste and lowering costs for residents. Smart city projects are already being implemented in places like Singapore, Tokyo, and Dubai, where AI is helping manage everything from public transportation to waste collection.

AI is also being used to improve public safety in smart cities. Surveillance systems powered by AI can detect suspicious behavior in real-time, alerting law enforcement to potential threats before they escalate. While this technology has the potential to make cities safer, it also raises significant concerns about privacy and surveillance. Who watches the AI, and how do we ensure it doesn't overreach?

Moreover, smart cities promise to improve the environment. AI-powered systems can manage water usage, reduce carbon emissions, and even monitor air quality. The potential for AI to help cities become more sustainable is enormous, but the success of these systems will depend on responsible design and implementation.

The Invisible Hand of AI: What We Don't See

While it's easy to point out the ways we see AI in our lives, much of AI's influence happens behind the scenes, in ways that are less obvious but equally transformative. Financial markets, for example, have been using AI-driven algorithms for years. These algorithms can make split-second trades based on vast amounts of data, and they often outperform human traders. AI is also optimizing supply chains, predicting demand fluctuations, and managing inventory in ways that would be impossible for human analysts alone.

Even in customer service, AI quietly manages interactions you may not realize. When you submit a query to a website's help center, an AI likely processes it, routes it to the right department, or provides an instant response. Many companies use AI to analyze customer feedback and complaints, identifying trends that help improve products and services.

In essence, AI is becoming the silent driver of efficiency in countless industries, often without us realizing the extent of its influence. As we continue to integrate AI into more areas of life, it's crucial to remain aware of its presence and think critically about how it's being used—both the good and the potential risks.

Conclusion: The AI Tapestry of Life

We're living in an era where AI is intricately woven into almost every aspect of modern life. From the personal assistants we carry in our pockets to the way we consume entertainment, work, and learn, AI is there, improving convenience and productivity at every step. The impact of AI on smart cities and behind-the-scenes industries shows just how widespread its influence has become.

As exciting as AI's potential is, these advancements come with challenges. Concerns about privacy, job displacement, and dependency on AI systems cannot be ignored. The balance between embracing AI's benefits and addressing its risks is delicate, but one that we must navigate carefully as we shape the future.

AI is no longer a futuristic concept—it's a reality, shaping the way we live, work, and interact with the world around us. From our personal devices to the corporate boardroom, AI has integrated itself into the fabric of modern life, offering convenience and efficiency at every turn. But with this rise comes significant challenges and questions. How will AI continue to shape industries? Will it be a job creator or destroyer? And what does it mean for the future of human intelligence and creativity?

As we move forward, one thing is clear: AI is here to stay. The next chapter will dive into the myths and fears surrounding AI, separating fact from fiction as we continue to explore the future we're building with this powerful technology.

Chapter 4: Myths and Misconceptions About AI

AI has become a hot topic in recent years, but with its rise in popularity comes a wave of myths, misconceptions, and misunderstandings. From dystopian visions of robots taking over the world to unrealistic expectations of AI's capabilities, public perception of AI often strays far from reality. In this chapter, we'll separate fact from fiction, addressing some of the most common myths surrounding AI.

Myth #1: AI Will Take Over the World

Let's start with the big one—AI is not going to take over the world, at least not in the way some people imagine. Sci-fi movies and books have long depicted AI as an all-powerful entity that will one day rise up against humanity, enslaving us or rendering us obsolete. While it's fun to think about scenarios where sentient machines turn on their creators, the reality is much less dramatic.

The AI we have today, known as narrow AI, is incredibly specialized. It can perform specific tasks very well, like recognizing images, translating languages, or playing chess, but it lacks the ability to think, reason, or make decisions the way humans do. In fact, AI has no consciousness, emotions, or desires—it's simply a tool programmed to process data and perform actions based on that data. The idea of machines developing their own will and deciding to overthrow humans is a far-fetched concept that belongs in Hollywood, not in reality.

That said, it's important to acknowledge that AI does raise some significant ethical questions about control and accountability. While AI may not be planning world domination, it can still cause harm if used irresponsibly, especially if it's deployed without proper oversight. The more pressing concern is not AI taking over, but humans misusing AI.

Myth #2: AI Is Already Smarter Than Humans

Another common misconception is that AI is smarter than humans. While it's true that AI can outperform humans in certain tasks, like calculations or data processing, it's not "smart" in the way people often think. AI systems can analyze vast amounts of data quickly and accurately, but they lack the ability to understand or interpret that data with any form of intuition or common sense.

Let's take an example: AI can identify faces in a photo with incredible accuracy, but it doesn't "know" who those people are or what they're feeling. It can recognize patterns, but it doesn't understand the context behind those patterns. This is a key difference between AI and human intelligence—AI is extremely good at narrow tasks but fails when it comes to general understanding and adaptability.

AI's seeming "intelligence" comes from its ability to process data and learn from it, but that doesn't mean it's capable of human-like reasoning. It's all about pattern recognition and statistical probability, not true cognitive thinking.

Myth #3: AI Will Take All of Our Jobs

Perhaps one of the most anxiety-inducing myths is the fear that AI will take over all human jobs, leading to widespread unemployment and economic disaster. While it's true that AI is automating certain tasks, and some jobs will inevitably be lost, the situation is not as bleak as it sounds.

History has shown that technological advancements often lead to job displacement, but they also create new opportunities. The Industrial Revolution, for example, eliminated many manual labor jobs, but it also gave rise to new industries and job roles that hadn't existed before. AI is following a similar pattern. Yes, some repetitive or low-skill jobs may be automated, but AI is also creating demand for new skills and jobs in areas like AI development, data science, and human-machine collaboration.

Moreover, many jobs—especially those that require creativity, emotional intelligence, and complex decision-making—are unlikely to be fully replaced by AI. Rather than eliminating all jobs, AI is more likely to change the nature of work, requiring people to adapt to new roles and skill sets.

Myth #4: AI Is Neutral and Unbiased

Another common myth is the belief that AI is inherently neutral and free from bias. After all, machines don't have personal beliefs or opinions, so how could they be biased? Unfortunately, the reality is that AI systems can—and often do—reflect the biases of the data they are trained on and the humans who build them.

AI learns from data, and if that data is biased, the AI will adopt those biases. For instance, if an AI system is trained on historical data that reflects discrimination in hiring practices, it may unintentionally learn to favor certain candidates over others based on factors like gender, race, or age. Similarly, AI systems used in criminal justice, healthcare, and other critical fields have been shown to perpetuate biases that can lead to unfair outcomes.

The problem isn't that AI is inherently biased, but that it reflects the biases present in the data it's fed. Addressing these biases requires careful attention to how AI systems are developed, as well as ongoing efforts to ensure that AI is used in ways that promote fairness and equality.

Myth #5: AI Will Solve All of Humanity's Problems

On the opposite end of the spectrum from the doomsday scenarios is the belief that AI is a silver bullet that will solve all of humanity's problems—from climate change to poverty to disease. While AI certainly has the potential to be a powerful tool in tackling global challenges, it's not a magical solution that can fix everything.

AI can help us analyze vast amounts of data, optimize systems, and make better decisions, but it's still limited by the quality of the data and the goals set by its human operators. More importantly, many of the world's most pressing issues—like inequality, political conflict, and environmental degradation—are deeply complex and cannot be solved by technology alone.

AI is a tool, not a panacea. It can amplify human efforts and provide valuable insights, but it's not going to single-handedly eliminate global challenges. The responsibility for addressing these issues still lies with human decision-makers and society as a whole.

Myth #6: AI Is Too Complex for the Average Person to Understand

There's a perception that AI is so complex that only experts in computer science and mathematics can truly grasp how it works. While AI is a highly technical field, understanding its basic concepts and implications is within reach for the average person. You don't need to know how to code or design an algorithm to understand what AI does and how it impacts your life.

In fact, one of the goals of this book is to demystify AI for everyday readers, breaking down complicated concepts into digestible ideas. By understanding how AI works at a basic level, you can make informed decisions about how to interact with it and how to advocate for its responsible use.

Conclusion: Separating Fact from Fiction

AI is often shrouded in myths and misunderstandings, which can lead to fear, confusion, or unrealistic expectations. While AI is a powerful and transformative tool, it's not the world-conquering, job-destroying, omnipotent force it's sometimes made out to be. At the same time, it's not a magic wand that can solve all of humanity's problems.

By dispelling these myths, we can have a more balanced and realistic conversation about AI—one that recognizes both its incredible potential and its limitations. It's essential that we approach AI with a nuanced understanding, recognizing that its success and impact depend on how we choose to develop and implement it. This means addressing the biases that can creep into AI systems, maintaining human oversight, and ensuring that AI benefits society as a whole, rather than just a select few.

As AI continues to evolve, it's up to us to shape its trajectory responsibly. A clear understanding of what AI can and cannot do will help us harness its strengths while mitigating its risks. In the next chapter, we'll move beyond the myths and explore the real-world impact of AI, diving into the ways it's already reshaping industries and economies worldwide.

Chapter 5: AI's Learning Curve

Artificial intelligence, or AI, is often perceived as a magical black box—a mysterious entity that effortlessly mimics human intelligence. However, the truth is far less enigmatic. AI operates on a learning curve that is both fascinating and complex, revealing the mechanisms through which these systems acquire knowledge, improve performance, and adapt over time. Understanding this learning curve is crucial for grasping the potential and limitations of AI in our lives.

The Basics of AI Learning

At its core, AI learning involves algorithms that process data to identify patterns and make predictions. The most common form of AI learning is known as machine learning, which can be broken down into several categories:

1. Supervised Learning: This involves training an AI model on a labeled dataset, where both the input data and the corresponding output are provided. For example, a supervised learning algorithm could be used to identify images of cats and dogs by learning from a dataset of labeled images. The algorithm makes predictions and is corrected by comparing its outputs to the known labels until it learns to make accurate classifications.

2. Unsupervised Learning: Unlike supervised learning, unsupervised learning uses data without explicit labels. The AI analyzes the input data to find hidden patterns or groupings. For example, it might discover that certain customers exhibit similar purchasing behaviors without prior knowledge of those behaviors. This is often used in market segmentation and anomaly detection.

3. Reinforcement Learning: This approach involves training AI agents through trial and error. The agent learns to make decisions by receiving feedback in the form of rewards or penalties based on its actions. For example, a reinforcement learning algorithm could be used to teach a robot to navigate a maze, where it receives positive feedback for reaching the exit and negative feedback for hitting walls.

These learning methods allow AI systems to improve their performance over time. The more data they are exposed to, the better they become at understanding complex patterns and making accurate predictions.

The Learning Process

AI's learning process can be likened to a child's education. Just as children learn by observing, practicing, and receiving feedback, AI systems also require time and experience to develop their capabilities. The journey of AI learning typically unfolds in several stages:

1. Data Collection: The foundation of any AI system is data. For machine learning algorithms to learn effectively, they require large amounts of high-quality data. This data can be sourced from various platforms, including social media, transactional records, or even user interactions. The quality of this data significantly impacts the AI's learning outcomes.

2. Training: Once the data is collected, it is used to train the AI model. This involves feeding the data into the algorithm, which then identifies patterns and relationships within the data. During this phase, the model adjusts its internal parameters to minimize errors and improve accuracy. It's akin to a student studying for an exam—repeated practice leads to better performance.

3. Validation and Testing: After training, the AI model undergoes validation and testing to assess its performance. This involves using a separate dataset that the model hasn't seen before to evaluate how well it can make predictions. This step is crucial for ensuring that the model generalizes well to new data rather than simply memorizing the training examples.

4. Deployment and Continuous Learning: Once validated, the AI model is deployed in real-world applications. However, the learning process doesn't end there. AI systems can continue to learn from new data and adapt to changing conditions. This ongoing learning enables them to improve performance over time, similar to how humans gain expertise through experience.

Challenges in the Learning Curve

While AI's learning curve presents significant opportunities, it also comes with challenges. One major hurdle is the data quality issue; poor-quality data can lead to inaccurate predictions and biased outcomes. AI systems can inadvertently perpetuate existing biases present in the training data, resulting in unfair or discriminatory practices.

Another challenge lies in the explainability of AI systems. As AI models grow more complex, understanding how they arrive at specific decisions becomes increasingly difficult. This lack of transparency can lead to mistrust among users, especially in high-stakes scenarios such as healthcare or criminal justice.

Additionally, the resource requirements for training AI systems can be substantial. High computational power and large datasets can be costly and may limit access for smaller organizations or individuals looking to leverage AI technology.

The Road Ahead

As we look to the future, the learning curve of AI will continue to evolve. Innovations in algorithms and data processing techniques promise to enhance the capabilities of AI systems, making them even more efficient and versatile. Moreover, as ethical considerations around AI use gain traction, the focus on developing transparent and fair AI systems will likely shape the trajectory of AI learning.

In conclusion, understanding AI's learning curve is essential for recognizing its potential and limitations. As we navigate this landscape, it is crucial to foster a balanced relationship with AI—one that embraces the benefits while remaining vigilant about the challenges. By doing so, we can ensure that AI serves as a valuable partner in our journey toward a more innovative and equitable future.

Chapter 5: AI's Learning Curve

PART TWO:

AI AND SOCIETY: SHAPING OUR WORLD

Chapter 6: AI in the Workplace – Reshaping Jobs and Industries

As AI technology continues to evolve, one of the most significant impacts is being felt in the workplace. From healthcare to manufacturing, finance to education, AI is changing how we work, the skills we need, and even the very nature of some industries. While some jobs are being automated, new opportunities are also emerging, requiring a shift in how we think about work in the 21st century.

In this chapter, we'll explore the ways AI is reshaping the workforce, the industries most affected by these changes, and what it means for the future of employment.

Automation: A Double-Edged Sword

The most immediate and visible impact of AI in the workplace is automation. Repetitive tasks—especially those that rely on data processing or routine decision-making—are increasingly being handled by AI systems. From factories using AI-powered robots to streamline production, to banks automating customer service with chatbots, the ability of AI to take over time-consuming tasks is transforming entire sectors.

For example, in manufacturing, AI-driven robots now perform tasks like assembling products, handling inventory, and quality control, all at a pace far beyond what humans could achieve. AI isn't just improving speed—it's also reducing error rates and operational costs. Similar changes are happening in industries like logistics, where AI helps optimize routes for delivery trucks and manage inventory with greater precision.

But automation, while boosting efficiency, is also a source of concern. Many fear that as AI takes over these tasks, jobs will disappear. In sectors like retail and customer service, automation of routine interactions through chatbots and AI-driven virtual assistants is already reducing the need for human workers in some roles. This is where the double-edged sword of AI becomes apparent: while it can help businesses grow and improve, it also threatens traditional jobs, especially those that rely on repetitive tasks.

New Jobs: Adapting to an AI-Driven Economy

Despite fears of mass job losses, history has shown that technological advancements often create new types of work. AI is no different. While it may automate certain jobs, it's also generating demand for new roles that didn't exist just a few years ago.

AI needs human oversight, maintenance, and development. As companies adopt AI technologies, they need skilled professionals to design, train, and manage AI systems. Data scientists, machine learning engineers, AI ethicists, and AI trainers are just a few of the new jobs emerging in the AI ecosystem. These roles require a combination of technical expertise, creativity, and an understanding of how AI systems operate and impact businesses.

Moreover, AI is transforming existing jobs rather than replacing them outright. For example, in fields like healthcare, AI assists doctors by analyzing medical data and offering diagnostic suggestions. However, the final judgment still rests with human physicians, who interpret the AI's findings and make complex decisions based on their knowledge and experience. The collaboration between AI and human professionals is becoming increasingly common in areas like law, finance, and education, where AI serves as a powerful tool to enhance human performance.

Rather than replacing humans, AI is augmenting our abilities, freeing up time to focus on higher-order tasks that require creativity, problem-solving, and emotional intelligence.

Industries Most Affected by AI

While AI is touching nearly every sector, some industries are feeling its impact more acutely than others. Here are a few examples of industries that are undergoing significant transformation due to AI:

Healthcare: AI is revolutionizing how we diagnose and treat diseases. Machine learning algorithms can analyze medical images, identify early signs of conditions like cancer, and suggest treatment options. In addition to diagnostics, AI is being used to predict patient outcomes, optimize hospital management, and even develop new drugs. However, the integration of AI in healthcare also raises ethical questions about privacy and the role of human decision-making.

Finance: The finance industry has long been a leader in AI adoption. Algorithmic trading, which relies on AI to make split-second decisions based on market data, now dominates global stock markets. AI is also used in fraud detection, credit scoring, and risk management, improving efficiency and accuracy. However, as AI takes on more responsibilities in managing investments and transactions, there are concerns about the potential for bias in AI-driven financial decisions.

Retail: In retail, AI is transforming both the customer experience and behind-the-scenes operations. Personalized recommendations, chatbots, and virtual shopping assistants are now common in online stores, while AI-powered supply chain management helps retailers optimize inventory and reduce costs. Brick-and-mortar stores are also experimenting with AI, using facial recognition for security and predictive analytics to tailor promotions to individual customers.

Manufacturing: AI is revolutionizing production lines, allowing for more efficient processes and quality control. Predictive maintenance, where AI anticipates machine failures before they occur, is saving manufacturers millions of dollars in downtime and repairs. As automation increases, factories are becoming smarter and more efficient, though this shift also means a reduced need for human labor in certain areas.

Media and Entertainment: AI is now involved in content creation, from generating news articles to composing music. AI-driven platforms like YouTube and Netflix use machine learning algorithms to recommend content to users based on their preferences. Even in journalism, AI is being used to automate the writing of basic news reports, raising questions about the future of creative professions.

The Future of Work: Collaboration Between Humans and AI

As AI continues to shape industries, the future of work will likely involve deeper collaboration between humans and machines. Instead of thinking of AI as a replacement for human workers, we need to view it as a tool that enhances our capabilities. AI can handle repetitive tasks, analyze data at lightning speed, and help make more informed decisions, but it still relies on human creativity, judgment, and ethical oversight.

The key to thriving in an AI-driven world is adaptability. Workers need to be open to learning new skills, particularly in areas that AI cannot easily replicate—creativity, critical thinking, problem-solving, and emotional intelligence. As AI continues to automate routine tasks, human workers will need to focus on what they do best: innovation, leadership, and interpersonal communication.

Education and training will play a critical role in preparing the workforce for the AI revolution. Governments, businesses, and educational institutions must work together to ensure that workers have access to the skills and knowledge they need to succeed in an AI-driven economy. Lifelong learning, reskilling, and upskilling will become essential as the nature of work evolves.

Conclusion: Embracing the Change

AI is reshaping the workforce in profound ways, automating tasks, creating new jobs, and transforming entire industries. While this shift presents challenges—particularly for workers in roles vulnerable to automation—it also offers tremendous opportunities for growth, innovation, and enhanced human-machine collaboration.

The future of work will not be about machines replacing humans, but about how we can use AI to complement our skills and amplify our potential. By embracing change and adapting to new roles and industries, we can ensure that AI serves as a force for positive transformation, both in the workplace and beyond.

At the same time, it's critical that we remain vigilant about the social and economic implications of AI in the workforce. Policymakers, businesses, and educators must collaborate to create a balanced environment where AI adoption does not leave entire groups behind, and where the benefits of AI are shared equitably. Supporting workers through reskilling initiatives and providing safety nets for those impacted by automation are essential steps to ensure a fair transition.

Ultimately, AI should be seen not as a threat, but as a tool to enhance human ingenuity and productivity. With the right framework, we can navigate the changes AI brings while preserving the value of human work, creativity, and purpose. As we move forward, the conversation should focus on how we shape a future where humans and AI work together in harmony, leveraging the best of both worlds.

Chapter 7: AI and the Global Economy– Redefining Growth and Power

AI is not just transforming workplaces and industries—it's reshaping the global economy. As businesses and governments alike invest heavily in AI technology, the economic landscape is shifting in unprecedented ways. Countries that lead in AI innovation are gaining new economic advantages, while those slow to adapt risk falling behind. The race for AI dominance is not just about technology—it's about who will control the future of economic growth, geopolitical power, and wealth distribution.

In this chapter, we will examine how AI is influencing the global economy, the power dynamics it is creating between nations, and how businesses are navigating the opportunities and challenges AI brings.

AI as an Economic Driver

AI is quickly becoming one of the most significant drivers of economic growth. Companies that successfully leverage AI can improve efficiency, reduce costs, and offer innovative products and services that were previously impossible. From automating supply chains to enhancing customer experiences with personalized marketing, AI is enabling

Chapter 7: AI and the Global Economy

businesses to scale faster and respond to market demands more effectively.

Tech giants like Google, Amazon, and Alibaba are at the forefront of AI research and application, using AI to optimize everything from their logistics operations to online advertising. These companies have realized that AI isn't just a tool—it's a strategic asset that can provide a competitive edge. As a result, they are investing billions of dollars in AI research and acquiring AI startups to stay ahead of the curve.

Beyond individual companies, AI is driving growth across entire economies. Countries that have embraced AI as part of their national strategy—such as the United States, China, and South Korea—are experiencing rapid technological advancement and economic growth. AI is becoming a key factor in determining which countries will be the economic leaders of the 21st century.

The Race for AI Dominance: Global Competition

The global race for AI dominance is well underway, and it's not just about technology—it's about power. Countries that lead in AI development and adoption will have a significant advantage in shaping the future of the global economy and geopolitics.

At the forefront of this race are the United States and China. Both countries are investing heavily in AI research and development, with ambitious national strategies aimed at becoming the world leader in AI by 2030. The U.S. benefits from its strong technology sector, home to some of the world's largest and most innovative tech companies. Silicon Valley remains a hub of AI research, attracting top talent from around the world.

China, however, is not far behind. With massive government investments, a booming tech industry, and access to vast amounts of data, China is positioning itself as a major AI powerhouse. Chinese companies like Baidu, Tencent, and Huawei are making significant strides in AI, and the country's government has made it clear that it views AI as critical to its economic and military future.

Other countries, such as Canada, South Korea, and the European Union, are also playing significant roles in the AI landscape. Canada, for example, has become a leader in AI research, particularly in the fields of deep learning and machine learning, while the EU is focusing on ethical AI development and ensuring that AI technologies are used responsibly.

The global race for AI dominance is not just about economic power—it's about who will set the rules for AI's use, development, and governance. The country that leads in AI will have the ability to shape international norms and regulations, giving it significant geopolitical influence.

AI's Impact on Global Wealth Distribution

AI's ability to transform industries and economies has profound implications for global wealth distribution. On one

Chapter 7: AI and the Global Economy

hand, AI could help reduce inequality by increasing productivity and creating new economic opportunities. By automating tasks, AI allows businesses to produce more with fewer resources, which could lower costs and make goods and services more accessible to a broader population.

However, there is also a risk that AI could exacerbate existing inequalities, both within countries and between them. Countries and companies that are at the forefront of AI development stand to gain the most from its economic benefits, while those that lag behind could face widening economic disparities. This is particularly concerning for developing nations, which may not have the resources or infrastructure to compete in the AI-driven global economy.

Within countries, there is a growing concern that AI could lead to a concentration of wealth in the hands of a few tech companies and their shareholders, while workers whose jobs are displaced by automation are left behind. Without proactive measures, AI could deepen the divide between the tech elite and the broader population, leading to social and economic instability.

Governments and policymakers will need to address these challenges by creating frameworks that ensure the benefits of AI are distributed more equitably. This might include investing in education and training programs to help workers adapt to the AI-driven economy, implementing tax reforms that encourage responsible AI development, and creating safety nets for those displaced by automation.

AI in International Trade and Commerce

AI is also reshaping international trade and commerce. AI-powered technologies are helping businesses streamline their global supply chains, optimize logistics, and predict market trends with greater accuracy. AI-driven predictive analytics, for example, can help companies anticipate changes in demand, allowing them to adjust production and distribution strategies accordingly.

In the world of finance, AI is revolutionizing how international transactions are conducted. AI-powered algorithms can detect fraudulent activities in real time, assess credit risk more accurately, and even manage complex financial portfolios. This has made global trade more efficient and secure, while also opening up new markets for businesses that previously lacked the resources to expand internationally.

At the same time, AI is introducing new challenges to the global trade landscape. Intellectual property rights, data privacy, and cybersecurity are all becoming critical issues as AI becomes more integrated into global commerce. Countries are grappling with how to regulate AI in a way that promotes innovation while protecting consumers and national security.

The Role of Governments and Policymakers

Governments have a crucial role to play in shaping the AI-driven global economy. They must balance the need to foster innovation with the responsibility to protect workers and ensure that AI's benefits are distributed fairly. This means investing in AI research and development, creating regulatory frameworks that address ethical concerns, and supporting reskilling initiatives to help workers transition to new jobs.

Moreover, governments need to collaborate internationally to establish global norms and regulations for AI. Issues like data privacy, algorithmic bias, and AI's role in national security require coordinated efforts across borders. The creation of international agreements on AI governance could help prevent a "race to the bottom" where countries compete by lowering standards to attract AI investments.

In the future, governments will need to ensure that AI is used not only as a tool for economic growth but also as a means of addressing global challenges such as climate change, poverty, and healthcare. AI has the potential to make a positive impact on these issues, but it will require careful planning and a commitment to using technology for the common good.

Conclusion: Navigating the AI-Driven Global Economy

AI is fundamentally altering the global economy, creating new opportunities for growth, innovation, and international collaboration. However, it also poses significant challenges, particularly when it comes to wealth distribution, job displacement, and global competition. As we move further into the AI era, countries, businesses, and individuals must be prepared to adapt to these changes and navigate the complexities of an AI-driven world.

The race for AI dominance is not just about technology—it's about who will shape the future of the global economy and wield influence in the 21st century. While AI offers incredible potential for economic growth, it's essential that we approach its development with caution, ensuring that its benefits are shared broadly and that its risks are mitigated through thoughtful governance and policy.

In the next chapter, we'll explore how AI is impacting society at large, from its influence on culture and education to the ethical dilemmas it raises. AI's reach goes beyond the workplace and the economy, touching nearly every aspect of our lives, and its societal implications are just as important to understand.

Chapter 8: AI and Society–Culture, Education, and Ethical Dilemmas

As AI becomes embedded in our daily lives, it's not just the economy or the workplace that's changing—society at large is being reshaped. AI is influencing our culture, transforming education, and presenting new ethical challenges that force us to reconsider the role of technology in human life. From the way we interact with information to the choices we make about how we live, AI is quietly but significantly altering the fabric of society.

In this chapter, we will explore how AI is influencing culture, how it's changing education, and the ethical dilemmas that arise from its widespread use.

AI and Culture: A New Digital Society

AI has infiltrated our cultural landscape in subtle yet profound ways. From the algorithms that recommend what we watch, read, or listen to, to AI-generated art and music, the cultural impact of AI is undeniable. Streaming platforms like Netflix and Spotify rely on AI to personalize content, creating tailored experiences that cater to individual preferences. But this algorithmic curation also raises questions: Are we losing the element of surprise or serendipity? Are we becoming more isolated in echo chambers of content that simply reinforce our existing tastes?

Social media platforms, powered by AI algorithms, shape how we communicate and engage with the world. AI curates our newsfeeds, showing us the information it thinks we'll engage with most. While this personalization can keep us entertained and informed, it can also create filter bubbles, limiting exposure to diverse perspectives and leading to polarization in society. In a world where AI decides what we see, are we losing our ability to critically engage with information?

AI's role in creative fields is also growing. AI-generated art, music, and even writing are becoming more common, with some AI-created works being sold in galleries or used in marketing campaigns. But this raises an important question: What does it mean to create? Can AI truly be considered creative, or is it merely replicating patterns it's been trained on? The debate over AI and creativity touches on broader questions about the role of human intuition and emotion in the creative process.

AI and Education: Rethinking Learning

AI is revolutionizing education, offering new tools for learning and reshaping the way we think about teaching. AI-driven platforms can tailor educational content to the needs of individual students, providing personalized learning experiences that adapt to different learning styles and paces.

Chapter 8: AI and Society—Culture, Education, and Ethical Dilemmas

Tools like AI tutors and grading systems are being implemented in classrooms around the world, offering students instant feedback and freeing up teachers to focus on more complex tasks.

One of the most significant benefits of AI in education is its ability to democratize access to information. AI-powered educational platforms are making high-quality education available to more people, especially in underserved regions. For example, AI can provide students in remote areas with access to the same resources as their peers in major cities, helping to close the education gap.

However, the integration of AI into education also presents challenges. The reliance on AI tools could lead to a depersonalization of education, where human interaction and mentorship take a backseat to machine-driven learning. Additionally, there are concerns about data privacy, as AI systems collect vast amounts of data on students' performance and behavior. Who controls this data, and how is it being used? And as AI takes on more of the teaching load, what happens to the role of educators? Will teachers become facilitators for AI, or will they continue to be central to the learning experience?

AI and Media: The Changing Landscape of Information

The media industry has experienced one of the most significant transformations due to AI. News organizations are increasingly relying on AI for content creation, data analysis, and even reporting. AI-generated news articles are becoming more common, especially in areas like financial reporting, sports, and breaking news, where speed and accuracy are crucial. AI systems can scan through vast datasets, identify trends, and generate coherent news stories in a fraction of the time it would take a human journalist. However, this efficiency raises concerns about the future of journalism. Will AI eventually replace human reporters, or will it serve as a tool to enhance their capabilities? And how will the rise of AI-generated content impact the quality and trustworthiness of information?

Moreover, the use of AI in the creation of deepfakes—manipulated videos or audio recordings that appear authentic—has further complicated the landscape of media and information. With the ability to create realistic but fake content, AI has introduced new challenges in the fight against misinformation and disinformation. In a world where it's becoming increasingly difficult to distinguish between what is real and what is fabricated, how do we ensure that AI is used responsibly in media? The growing prevalence of AI-generated content demands new ethical guidelines and robust mechanisms to maintain the integrity of information.

The Human-AI Interaction: Friend or Foe?

As AI becomes more integrated into our lives, the nature of human-AI interaction is evolving. From virtual assistants like Siri and Alexa to customer service bots, AI is increasingly becoming a primary point of contact in our daily interactions.

Chapter 8: AI and Society—Culture, Education, and Ethical Dilemmas

While AI can make these interactions more convenient and efficient, it also raises questions about the quality of human connection in a world mediated by machines. Are we trading meaningful human interactions for the ease and convenience of AI? And how does this shift impact our social skills, empathy, and ability to relate to one another?

The balance between leveraging AI's capabilities and preserving authentic human connections will be a defining challenge for society moving forward.

The Ethical Dilemmas of AI

The rise of AI has brought with it a host of ethical concerns. From privacy issues to algorithmic bias, AI's integration into society is raising questions about fairness, transparency, and accountability. One of the most pressing ethical challenges is the issue of bias in AI systems. Because AI algorithms are trained on historical data, they can sometimes reflect and amplify the biases present in that data, leading to unfair outcomes. For example, AI systems used in hiring or law enforcement have been shown to disproportionately disadvantage certain demographic groups, raising concerns about discrimination.

AI also presents ethical challenges around data privacy. AI systems rely on vast amounts of data to function effectively, but this data often comes from users without their explicit consent or knowledge. The rise of AI-powered surveillance systems, both online and in public spaces, has led to concerns about the erosion of privacy. As AI becomes more powerful, how do we ensure that individuals' rights to privacy are protected?

Another ethical issue is the potential for AI to be used in harmful ways. AI technologies have the potential to be weaponized, whether through autonomous military systems or malicious cyber-attacks.

Finally, there is the question of responsibility. As AI systems become more autonomous, who is responsible for their actions? If an AI system makes a mistake—whether it's in diagnosing a patient, approving a loan, or making a legal decision—who is held accountable? These questions are becoming more urgent as AI systems take on more decision-making roles in society.

Navigating Ethical Dilemmas in AI Implementation

As AI technology becomes more integrated into society, ethical dilemmas surrounding its use become increasingly complex and urgent. For instance, the deployment of AI in law enforcement raises significant concerns regarding surveillance and profiling, potentially leading to biased outcomes that disproportionately affect marginalized communities. The algorithms used to predict criminal behavior can inadvertently perpetuate existing prejudices if they are trained on biased data. Similarly, in the realm of education, the use of AI tools for personalized learning can raise questions about data privacy and consent, especially when children are involved. Ensuring that AI systems are transparent, fair, and accountable is paramount to maintaining public trust and protecting individual rights. Addressing these ethical dilemmas requires collaboration among technologists, ethicists, educators, and policymakers, creating a multidisciplinary approach to crafting guidelines and regulations that govern AI use. As we navigate the intricacies of AI's societal impact, fostering an ethical framework for its implementation will be crucial to harnessing its benefits while mitigating potential harms.

Conclusion: AI's Double-Edged Sword in Society

AI's impact on society is multifaceted. On one hand, it has the potential to revolutionize how we live, learn, and create, offering incredible benefits in terms of efficiency, access, and innovation. On the other hand, it presents significant ethical and social challenges that must be addressed to ensure that AI serves the common good.

As AI continues to evolve, we must navigate these complexities with care, balancing the opportunities it offers with the responsibilities it entails. AI is not a neutral tool—it reflects the values and biases of the society that creates it. By engaging in thoughtful discussions about how AI is integrated into our culture, education, and ethics, we can shape a future where AI enhances human life, rather than detracts from it.

Chapter 9: AI and the Workforce – Disruption or Transformation?

As AI continues to evolve, its impact on the workforce is one of the most widely debated topics. From automation to AI-powered decision-making, the way we work is being fundamentally transformed. Some view this as an exciting opportunity for innovation and increased efficiency, while others worry about mass unemployment and the loss of human skills. In this chapter, we'll explore how AI is reshaping the workforce, what jobs are at risk, and how industries and individuals can adapt to this rapidly changing landscape.

Automation and Job Displacement: What's at Stake?

One of the biggest fears surrounding AI is the displacement of jobs. AI-powered automation has already begun to replace human labor in industries ranging from manufacturing and retail to transportation and customer service. Self-checkout machines, AI-powered logistics systems, and even autonomous vehicles are just a few examples of how automation is reducing the need for human workers in traditionally labor-intensive roles.

A report by the World Economic Forum predicted that by 2025, machines will perform more tasks than humans in the workplace. Jobs that involve repetitive tasks, data processing, and routine decision-making are particularly vulnerable to automation. For instance, roles in manufacturing, data entry, and basic administrative work are being automated at an alarming rate. Even more complex tasks, like legal research or medical diagnostics, are increasingly being handled by AI systems. With AI's ability to learn and improve its processes, these technologies are becoming more capable, leaving many workers to wonder if their roles will eventually become obsolete.

However, the conversation about AI and jobs is not solely one of loss. While some jobs will disappear, new ones will emerge. The rise of AI has created demand for skills in machine learning, data science, and AI ethics, areas where human oversight and creativity are still essential. Moreover, jobs that require emotional intelligence, empathy, and complex problem-solving remain less vulnerable to automation, at least for now. As AI continues to evolve, the key will be adaptability—learning how to work alongside AI and leverage its capabilities to enhance human potential.

AI as a Tool for Productivity and Innovation

While AI poses a threat to certain jobs, it also offers tremendous opportunities to enhance productivity and drive innovation. AI can handle vast amounts of data in real-time, providing businesses with insights and solutions that would be impossible for humans to process on their own. In fields like finance, healthcare, and marketing, AI is enabling companies to operate more efficiently, make better decisions, and innovate at a faster pace.

Take, for example, the use of AI in legal services. AI-powered tools can now draft contracts, analyze legal documents, and even predict case outcomes based on past rulings. This allows law firms to streamline processes, reduce costs, and focus their human lawyers on more complex, strategic work. In healthcare, AI is revolutionizing diagnostics by analyzing medical images, identifying patterns, and predicting disease progression. By assisting doctors with data-driven insights, AI enhances accuracy and efficiency, leading to better patient outcomes.

In creative fields, AI is also proving to be a valuable tool. AI-generated music, design, and even writing are becoming more common, allowing creators to experiment with new forms and ideas. AI doesn't replace the creative process, but it can act as a powerful tool that enhances human creativity by automating certain tasks and suggesting novel approaches.

Ultimately, AI has the potential to transform the way we work—not by eliminating human roles, but by enabling us to focus on higher-level tasks that require creativity, empathy, and critical thinking. The future of work may involve fewer routine jobs, but it will also involve more collaboration between humans and machines, with AI serving as a tool that amplifies human abilities.

Chapter 9: AI and the Workforce – Disruption or Transformation?

The Changing Skillset: Adapting to the AI Revolution

The integration of AI into the workforce demands a shift in the skills required to succeed in the future. Technical skills like programming, data analysis, and machine learning are becoming increasingly valuable as businesses across industries adopt AI technologies. But equally important are soft skills like critical thinking, problem-solving, and emotional intelligence—areas where AI cannot (yet) compete with human abilities.

In fact, as AI takes over more routine tasks, the value of uniquely human skills is likely to increase. Jobs that require deep interpersonal interaction, such as nursing, counseling, and customer service, will still demand a human touch. Similarly, roles that require creative problem-solving and innovation, such as product design or entrepreneurship, are less likely to be fully automated.

As the workforce adapts to this new reality, lifelong learning will become essential. Workers will need to continually update their skills, not only in technical areas but also in developing their ability to work alongside AI. Education systems must also evolve to prepare students for a world where AI plays a central role, teaching them both how to work with AI and how to develop the critical thinking and emotional intelligence that machines cannot replicate.

Governments, too, have a role to play in facilitating this transition. Policies that promote upskilling, retraining, and social safety nets will be crucial in ensuring that workers displaced by AI are able to find new opportunities in the evolving economy.

AI and the Gig Economy

Another significant area of impact is the gig economy. AI is not just transforming traditional workplaces but also changing how freelancers and gig workers find and perform jobs. Platforms that match workers to gigs—such as ride-sharing apps or freelance marketplaces—are increasingly powered by AI algorithms that optimize job matching, predict demand, and streamline services. While this creates more opportunities for gig workers, it also raises concerns about job security, worker rights, and the power of algorithms in determining access to work.

Gig workers may find themselves increasingly dependent on AI-driven platforms to earn a living, with little control over the terms and conditions of their work. This shift highlights the need for regulatory frameworks that protect the rights of workers in the gig economy, ensuring fair wages, job stability, and protection from algorithmic bias.

The Rise of the Hybrid Workforce

As the integration of AI into various sectors continues to evolve, a new model of work is emerging: the hybrid workforce. This model combines human intelligence with AI capabilities, creating a symbiotic relationship where both can thrive. In industries like healthcare, education, and

Chapter 9: AI and the Workforce – Disruption or Transformation?

technology, professionals are increasingly collaborating with AI tools to enhance their productivity and creativity. For instance, doctors are using AI systems to analyze patient data and recommend treatment options, allowing them to spend more time on patient care rather than administrative tasks. Similarly, educators are employing AI-driven platforms to personalize learning experiences for students, tailoring educational content to individual needs. This shift is not just about augmenting human labor but also about redefining roles and responsibilities to harness the strengths of both humans and machines.

However, the hybrid workforce model also brings challenges that need to be addressed. As employees work alongside AI, there is a growing need for continuous learning and upskilling to keep pace with technological advancements. Organizations must invest in training programs that empower their workforce to adapt to new tools and workflows effectively. Furthermore, fostering a culture of collaboration between human workers and AI systems is essential to maximize the benefits of this new working dynamic. Companies must encourage open communication and provide opportunities for employees to share their insights and experiences, ensuring that human perspectives remain at the forefront of decision-making processes. Embracing the hybrid workforce approach not only enhances productivity but also helps create a more resilient and adaptable workforce prepared to face the challenges of the future.

The Role of Employers in Transitioning to an AI-Enhanced Workforce

As AI continues to reshape the workforce, employers have a crucial role to play in facilitating a smooth transition for their employees. This involves not only investing in new technologies but also ensuring that their workforce is equipped with the necessary skills to thrive in an AI-driven environment. Organizations must prioritize reskilling and upskilling initiatives that empower workers to embrace new roles that AI creates while minimizing the risks of job displacement. By fostering a culture of lifelong learning and adaptability, employers can help their teams navigate the complexities of an AI-enhanced workplace.

Additionally, companies should actively engage in discussions around ethical AI use and its implications for their workforce. Employers need to establish clear policies on how AI technologies will be integrated into their operations and ensure transparency in their implementation. This includes addressing concerns about job security, privacy, and potential biases in AI systems. By fostering an inclusive environment where employees feel informed and valued in the AI transition process, organizations can build trust and support among their teams. Ultimately, the success of integrating AI into the workplace hinges not only on technological advancements but also on the commitment of employers to invest in their most valuable asset—their people.

Chapter 9: AI and the Workforce – Disruption or Transformation?

Ethics and Regulation: Who Governs AI in the Workforce?

The use of AI in the workplace raises significant ethical and regulatory concerns. As AI systems become more involved in decision-making processes—whether it's in hiring, performance evaluation, or resource allocation—questions about bias, transparency, and accountability become increasingly important. AI systems are only as good as the data they are trained on, and if that data contains biases, the AI will perpetuate those biases in its decision-making.

For example, AI algorithms used in hiring have been criticized for discriminating against certain groups based on gender, race, or socioeconomic background. These systems, designed to be efficient and objective, can unintentionally reinforce existing inequalities if not properly monitored and audited. In response, there is a growing movement to develop ethical guidelines and regulatory frameworks for the use of AI in the workplace. Ensuring that AI is transparent, accountable, and fair will be critical to fostering trust in these technologies and protecting workers' rights.

Another concern is the potential for AI to erode workers' privacy. AI systems can track employees' activities, monitor their productivity, and even predict their behavior. While this can help companies optimize performance, it also raises concerns about surveillance and the balance between efficiency and privacy. How much data should companies be allowed to collect, and how should it be used? As AI becomes more ingrained in the workplace, these ethical questions will need to be addressed to ensure that AI serves the interests of workers as well as employers.

Conclusion: Navigating the Future of Work

AI is undoubtedly transforming the workforce, presenting both challenges and opportunities. While some jobs will be lost to automation, others will be created, and the nature of work itself will change. Rather than viewing AI as a threat, we should see it as a tool for enhancing human potential. By developing the skills necessary to work alongside AI and addressing the ethical concerns that come with its use, we can create a future where AI empowers rather than replaces us.

The key to navigating this future is adaptability. Workers, businesses, and governments alike must be prepared to embrace change, invest in education and training, and ensure that AI is used in ways that benefit society as a whole. As AI continues to reshape the world of work, those who are able to adapt will find new opportunities in an economy transformed by intelligent machines.

Chapter 10: AI and Governance – Regulating the Future

As artificial intelligence continues to integrate into nearly every aspect of society, the question of how to regulate its use becomes more pressing. Governments, policymakers, and industry leaders are grappling with the complexities of creating regulations that foster innovation while also safeguarding public interests. The rapid development of AI has outpaced the legislative process, leaving many governments playing catch-up. But in this race, the stakes are high—privacy, security, fairness, and ethical concerns must all be addressed. This chapter explores the current landscape of AI regulation, the challenges that governments face, and the international efforts being made to set guidelines for AI's development and deployment.

The Need for AI Regulation

As AI becomes more powerful and pervasive, the need for regulation has become clear. Without oversight, AI systems could perpetuate biases, infringe on privacy, or even pose risks to national security. While AI offers tremendous potential for societal good—such as improving healthcare, optimizing energy use, or enhancing public safety—it also carries risks. These risks include unintended consequences of autonomous decision-making, the amplification of existing social inequalities, and the potential for misuse in areas like surveillance or warfare.

Governments are beginning to recognize the importance of addressing these issues, but regulation presents a delicate balancing act. Too much regulation could stifle innovation, while too little could leave societies vulnerable to the negative impacts of unchecked AI use. A thoughtful, measured approach to AI regulation is necessary to strike this balance.

Global Approaches to AI Governance

Different countries have adopted various approaches to AI governance, reflecting their unique priorities and political climates. For instance, the European Union has taken a strong stance on AI ethics and data privacy, most notably with its General Data Protection Regulation (GDPR), which sets strict guidelines on how companies collect and process personal data. The EU is also working on an AI Act that aims to regulate high-risk AI systems in critical sectors like healthcare, law enforcement, and transportation.

In contrast, the United States has adopted a more laissez-faire approach, focusing on encouraging innovation and allowing companies to self-regulate. The U.S. government has issued AI-related guidance but has yet to implement comprehensive, binding regulations on AI use. China, meanwhile, has rapidly advanced in AI development and is integrating AI into state functions, including surveillance and governance. China's model of centralized control allows for quicker implementation of AI technologies but raises concerns about human rights and the misuse of AI for authoritarian purposes.

Ethical AI Frameworks and Industry Standards

Beyond government regulations, industries are also developing their own ethical AI frameworks. Major technology companies like Google, Microsoft, and IBM have introduced AI principles that emphasize fairness, transparency, and accountability. These principles aim to guide the development of AI in a way that benefits society, but they also raise questions about self-regulation and whether companies can be trusted to police themselves.

In many cases, these ethical frameworks are voluntary and lack the enforcement mechanisms needed to ensure compliance. This is where government intervention becomes crucial. To be effective, AI governance requires a combination of regulatory oversight, industry cooperation, and public involvement. It is essential to establish clear rules for how AI should be developed and used, particularly in high-stakes sectors like healthcare, finance, and criminal justice.

AI and National Security

The intersection of AI and national security is another critical area of concern. AI technologies are being increasingly integrated into defense systems, from autonomous drones to cybersecurity tools that protect against digital threats. While these technologies offer significant advantages, they also introduce new risks. Autonomous weapons, for example, could make decisions without human oversight, potentially leading to unintended escalations in conflict. Furthermore, AI-driven cyberattacks could target critical infrastructure, causing widespread disruption.

International collaboration on AI governance is crucial to addressing these concerns. Agreements like the UN's Convention on Certain Conventional Weapons (CCW) are beginning to address the ethical and practical challenges posed by autonomous weapons, but more work is needed. Governments must collaborate not only to establish rules for AI use in warfare but also to prevent an AI arms race that could destabilize global security.

The Role of AI in Policymaking

AI is not only a subject of governance but also a tool for governance. Governments are increasingly using AI to improve decision-making processes, streamline public services, and manage complex data. AI can help optimize traffic systems, allocate resources in healthcare, and even predict natural disasters. In policymaking, AI tools can analyze vast amounts of information, providing insights that help shape more effective policies.

However, the use of AI in governance also raises concerns about transparency and accountability. When AI systems are involved in making decisions that affect citizens, such as determining social benefits or assessing risk in criminal justice, it is essential that these systems are transparent and their decisions explainable. The public must have confidence that AI is being used fairly and that there are mechanisms in place to correct errors or biases in AI-driven decisions.

Challenges to AI Regulation

Regulating AI is an unprecedented challenge because of its complexity and rapid evolution. One major hurdle is the technical nature of AI, which requires lawmakers to have a deep understanding of the technology. This knowledge gap makes it difficult for legislators to draft effective laws that address the nuanced issues of AI use. Additionally, the global nature of AI development means that regulations in one country can have far-reaching effects in others, further complicating the process of creating consistent governance frameworks.

Another challenge is the sheer diversity of AI applications, from consumer technology to military defense, making it difficult to develop a one-size-fits-all regulatory approach. What works for regulating AI in healthcare might not be suitable for governing its use in finance or law enforcement. Policymakers must adopt a flexible, adaptive approach to regulation that can evolve alongside AI technology.

The Role of Public Opinion in AI Regulation

An often overlooked aspect of AI governance is the role of public opinion in shaping regulation. The public's understanding and perception of AI technology heavily influence the direction of policy debates. When people fear job loss, surveillance, or privacy violations, these concerns tend to drive calls for more stringent regulations. However, the general public's lack of technical understanding can sometimes lead to misunderstandings about AI's capabilities and risks. Governments and organizations must focus on improving AI literacy, ensuring that citizens are informed about both the benefits and potential dangers of AI. This empowers the public to engage meaningfully in discussions about AI policy, rather than reacting to sensationalized narratives.

International Collaboration: The Key to Unified Governance

One of the biggest challenges in AI regulation is its inherently global nature. AI systems developed in one country can be deployed across borders, affecting citizens worldwide. This makes international collaboration essential for creating unified standards and preventing regulatory gaps that could be exploited by bad actors. There are currently efforts underway to create global AI governance frameworks, but there is still a long way to go before we see a cohesive global strategy.

Conclusion: The Future of AI Governance

The future of AI governance will depend on how well governments, industries, and civil society can work together to create frameworks that protect the public while encouraging innovation. AI has the potential to transform industries, improve quality of life, and solve some of the world's most pressing problems—but only if it is developed and deployed responsibly.

In the coming years, we are likely to see more collaborative efforts to establish global norms for AI governance, with an emphasis on ethics, transparency, and accountability. The challenge will be to ensure that these efforts keep pace with the technology itself, ensuring that AI benefits everyone and does not exacerbate inequalities or create new risks.

AI is not just a technological challenge—it is a governance challenge, and how we navigate this complex terrain will shape the future of society.

Chapter 11: AI and Privacy – Navigating the New Data Landscape

In a world where artificial intelligence increasingly permeates everyday life, the topic of privacy is becoming more critical than ever. AI systems thrive on data—our data. They learn from our behavior, track our preferences, and predict our next moves, often with astounding accuracy. But as AI becomes more integrated into personal devices, online platforms, and even public infrastructure, questions about how our data is being collected, used, and protected take center stage. This chapter delves into the complex relationship between AI and privacy, exploring how data powers AI, the risks associated with data misuse, and the regulatory efforts being made to protect individuals in this new data landscape.

AI's Appetite for Data

Artificial intelligence, particularly machine learning models, relies on vast amounts of data to function effectively. The more data an AI system has access to, the better it can "learn" and make accurate predictions. For instance, AI-driven recommendation systems on platforms like Netflix or YouTube analyze our viewing habits to suggest content tailored to our tastes. While this can enhance user experience, it also means that our personal data is being continuously

Chapter 11: AI and Privacy – Navigating the New Data Landscape

collected, stored, and analyzed, often without us fully understanding the extent of it.

In addition to personal data from social media platforms, search engines, and e-commerce sites, AI systems also utilize data from smart devices, health records, and even public surveillance. This abundance of data allows AI systems to make more informed decisions, but it also raises significant concerns about the amount of control individuals have over their personal information. How much of our data is being collected? Who is using it? And for what purposes?

The Erosion of Privacy in the Age of AI

One of the biggest concerns regarding AI is its potential to erode personal privacy. With AI-powered systems increasingly embedded in our daily lives—from smart speakers in our homes to facial recognition software in public spaces—our every move, preference, and interaction can be tracked. This constant surveillance creates an environment where individuals have little control over their own data, making privacy a luxury rather than a right.

AI-driven surveillance technologies are particularly troubling. Governments and private companies alike are utilizing AI to monitor public spaces, often justifying this practice in the name of security or efficiency. For example, facial recognition technology is being used to track individuals in airports, shopping centers, and even on city streets. While this technology can improve public safety, it also poses significant risks to civil liberties, as it can be used to monitor citizens without their consent or knowledge. In countries with fewer protections on privacy rights, these technologies can even be weaponized for political control or oppression.

AI and Data Security: A Fragile Balance

The vast amounts of data used to train AI systems also make them a prime target for cyberattacks. If AI systems are built on sensitive personal information, the consequences of a data breach can be severe. For instance, a breach involving AI healthcare systems could expose patients' medical records, or a hack targeting financial institutions could reveal detailed banking information. With more data being fed into AI systems, there is an increased need for robust cybersecurity measures to protect against these risks.

Moreover, the interconnectedness of AI systems means that a vulnerability in one part of the system can have widespread effects. For example, AI-powered smart cities that manage everything from traffic flow to public utilities could be vulnerable to cyberattacks, leading to potential disruptions in essential services. The challenge for developers and policymakers is to ensure that AI systems are designed with strong security measures from the outset, rather than addressing vulnerabilities after they have been exploited.

Regulatory Frameworks: Protecting Privacy in the AI Era

In response to growing concerns over data privacy, governments around the world have started implementing regulations aimed at protecting individuals' personal information. The most notable of these is the European Union's General Data Protection Regulation (GDPR), which sets strict guidelines for how companies collect, store, and use personal data. Under the GDPR, individuals have the right to know what data is being collected about them, how it is being used, and can request that their data be deleted.

Other countries have followed suit, introducing their own versions of data protection laws. In the United States, states like California have implemented legislation such as the California Consumer Privacy Act (CCPA), which gives consumers greater control over their personal information. However, global consistency in data privacy laws remains a challenge. As AI technologies continue to evolve, it is essential that regulations keep pace to protect individuals' rights in this fast-changing environment.

The complexity of AI's reliance on data means that regulation must strike a delicate balance between encouraging innovation and protecting privacy. While comprehensive regulations like the GDPR provide important safeguards, they also introduce challenges for companies that rely on large-scale data processing. Ensuring that privacy protections are both effective and practical will require ongoing collaboration between governments, technology companies, and civil society.

The Rise of Privacy-Enhancing Technologies (PETs)

As concerns over privacy grow, the development and use of Privacy-Enhancing Technologies (PETs) have become crucial in mitigating the risks associated with AI-driven data collection. These technologies aim to protect personal data while allowing AI systems to function effectively. For instance, techniques like differential privacy allow AI to learn from datasets without directly accessing individual information. Federated learning is another method where data remains localized, and AI models are trained across multiple devices, thus preserving the privacy of user data. These advancements show that it is possible to balance the need for vast amounts of data in AI with the fundamental right to privacy.

Despite these innovations, the adoption of PETs is still in its early stages. Many companies and organizations have yet to fully implement these technologies, either due to a lack of understanding, resources, or urgency. As AI continues to evolve, it will be essential for businesses to adopt PETs not only to comply with regulatory requirements but also to build trust with consumers. By investing in privacy-conscious solutions, companies can demonstrate a commitment to safeguarding personal data, which can ultimately lead to greater acceptance and success for AI systems in the long run.

Chapter 11: AI and Privacy – Navigating the New Data Landscape

Empowering Individuals: Data Ownership and Control

An emerging discussion in the realm of AI and privacy is the concept of data ownership. In many cases, individuals are unaware of how much data they are sharing or have little control over how it is used once collected. This has sparked debates about the need for more robust frameworks that give people greater control over their personal information. Concepts like "data dignity" and "data sovereignty" suggest that individuals should be treated as the owners of their data, with the ability to decide how it is shared and monetized.

In addition to regulatory and technological solutions, education will play a vital role in ensuring individuals can make informed decisions about their data. Public awareness campaigns, digital literacy initiatives, and transparency from AI developers can help people understand the implications of sharing personal information and enable them to make more privacy-conscious choices.

Ethical AI: Building Trust through Transparency

Beyond regulatory frameworks, one of the most important aspects of protecting privacy in the age of AI is building systems that prioritize transparency and accountability. Many people feel that they have lost control over their personal data, with AI systems operating as "black boxes" where decisions are made without any clear explanation of how or why. This lack of transparency creates distrust between AI developers and users, exacerbating concerns about privacy.

To address this, AI developers are increasingly focusing on creating ethical AI systems that are designed to be explainable, transparent, and accountable. This means making the algorithms and data sources used by AI systems accessible and understandable to users, and ensuring that individuals have control over their data.

Conclusion: Charting a Privacy-Conscious Future for AI

The relationship between AI and privacy is complex, and the stakes are high. While AI has the potential to drive innovation and improve our lives, it also poses serious risks to individual privacy if left unchecked. Striking a balance between harnessing the benefits of AI and protecting personal data is one of the greatest challenges of the digital age.

Governments, companies, and individuals must work together to create a future where AI is used responsibly and ethically. This means developing robust regulatory frameworks, building AI systems that prioritize transparency, and fostering a culture of accountability. The goal should not be to halt the progress of AI but to ensure that its development aligns with fundamental human rights and values. As we continue to integrate AI into more aspects of our lives, the question is not whether AI will transform our society—but how we can ensure that this transformation respects our privacy and autonomy.

Chapter 11: AI and Privacy – Navigating the New Data Landscape

PART THREE:

HUMANITY'S ROLE IN AN AI DRIVEN WORLD

Chapter 12: AI and Human Rights

Artificial intelligence is not just a technological innovation; it is a force that could redefine fundamental human rights. As AI becomes more intertwined with daily life, it brings both opportunities and risks for human rights. The promise of AI in improving healthcare, education, and access to information is immense, but so too are the potential threats it poses to privacy, equality, freedom of expression, and even personal autonomy. In this chapter, we will explore how AI intersects with human rights, the ethical dilemmas it raises, and how we can ensure that the development of AI protects and promotes human dignity.

AI and Privacy: The Right to Be Left Alone

One of the most immediate and pressing concerns about AI is its impact on privacy. AI systems can collect and process vast amounts of personal data, often without the knowledge or consent of the individuals involved. From social media algorithms to facial recognition systems used in public spaces, AI is being deployed to monitor and predict human behavior on an unprecedented scale.

This level of surveillance raises significant questions about the right to privacy. In a world where AI can track individuals' movements, communications, and even emotions, the very concept of privacy could be at risk. While AI can improve security and efficiency, it also has the potential to create a surveillance state where individuals lose control over their personal information. The challenge is finding a balance between the benefits of AI and the protection of privacy rights, ensuring that individuals retain control over their data and how it is used.

Algorithmic Bias and the Right to Equality

Another critical human rights issue in the AI era is algorithmic bias. AI systems are often trained on large datasets, which reflect the biases and inequalities present in society. As a result, AI can inadvertently perpetuate and even amplify discrimination based on race, gender, religion, or socioeconomic status. For example, AI-powered recruitment tools have been found to favor male candidates over female ones, and facial recognition systems have been shown to be less accurate for people with darker skin tones.

The right to equality is enshrined in international human rights law, yet AI can undermine this principle when not designed or implemented properly. Ensuring that AI systems are fair, transparent, and free from bias is essential to upholding the right to equality. This requires not only technical solutions, such as improving the diversity of training data, but also legal frameworks that hold AI developers accountable for discriminatory outcomes.

Chapter 12: AI and Human Rights

Freedom of Expression and the Role of AI in Information Control

AI plays a significant role in shaping the information that individuals receive, particularly through social media platforms and search engines. These systems use AI algorithms to determine which content is most relevant to users, often prioritizing sensational or controversial material that generates more engagement. While this can enhance user experience by delivering personalized content, it also raises concerns about freedom of expression and the manipulation of information.

AI-driven content moderation can lead to censorship, as algorithms may block or de-prioritize certain viewpoints without human oversight. In authoritarian regimes, AI is already being used to suppress dissent and control the flow of information, posing a direct threat to freedom of speech and access to information. The challenge here is ensuring that AI systems do not stifle freedom of expression while still addressing legitimate concerns about harmful or misleading content.

AI in Law Enforcement: Threats to Freedom and Justice

AI is increasingly being used in law enforcement, from predictive policing algorithms to facial recognition technology. While these tools can enhance public safety, they also pose significant risks to human rights, particularly the rights to freedom, justice, and due process. Predictive policing, for instance, uses historical crime data to forecast where future crimes are likely to occur. However, these algorithms are often based on biased data, leading to the over-policing of marginalized communities.

Similarly, AI-driven facial recognition can lead to wrongful arrests, as the technology is not always accurate, particularly when identifying people of color. The use of AI in law enforcement raises concerns about accountability and transparency, as individuals may find it difficult to challenge decisions made by opaque algorithms. Ensuring that AI systems used in law enforcement respect human rights will require robust legal safeguards and oversight mechanisms to prevent abuses of power.

The Right to Work in an AI-Driven World

As discussed in earlier chapters, AI is transforming the workplace, with both positive and negative implications for the right to work. While AI can enhance productivity and create new job opportunities, it also threatens to displace workers in many industries. The right to work is recognized as a fundamental human right, and ensuring that AI development does not lead to mass unemployment is a critical challenge for policymakers.

The future of work in an AI-driven world will depend on how well societies can manage this transition, ensuring that everyone has access to meaningful employment.

Conclusion: Ensuring Human Rights in an AI-Driven Future

As AI continues to evolve, it will have profound implications for human rights. While AI offers the potential to enhance human dignity by improving access to services and creating new opportunities, it also poses significant risks to privacy, equality, freedom, and autonomy. The challenge for policymakers, technologists, and society at large is to ensure that AI is developed and deployed in ways that protect and promote human rights, rather than undermine them.

Governments must establish clear legal frameworks that address the human rights implications of AI, while businesses must prioritize ethical AI development. By placing human rights at the center of AI governance, we can ensure that AI serves the public good, rather than becoming a tool of oppression or exploitation.

Chapter 13: Coexisting with Machines Finding Balance in an AI-Driven World

As artificial intelligence advances and integrates into almost every facet of human life, a crucial question arises: how do we coexist with machines? AI has moved from the realm of science fiction to our homes, workplaces, and even our decision-making processes. From virtual assistants and self-driving cars to AI-powered medical diagnostics and personalized recommendations, machines are becoming indispensable in our daily routines. But as we grow more reliant on AI, the challenge is finding a balance—one that allows us to harness the power of machines while preserving what makes us distinctly human.

In this chapter, we explore the dynamics of coexisting with AI-driven machines, how to navigate the relationship between humans and technology, and how to shape a future where machines augment, rather than dominate, human existence.

The Rise of Intelligent Machines in Everyday Life

AI is no longer confined to research labs or high-tech industries—it's now a part of the everyday world. Voice-activated assistants like Siri, Alexa, and Google Assistant respond to our commands; algorithms on social media tailor our newsfeeds; and AI-powered chatbots handle customer service inquiries. We are already sharing our lives with machines, whether we realize it or not.

One of the most significant impacts of AI on daily life is its ability to automate tasks, simplifying complex processes that once required human effort. From optimizing traffic flow to recommending movies, machines are taking on roles that free us from mundane chores. However, as machines perform more tasks previously done by humans, we face new dilemmas. Where do we draw the line between convenience and over-reliance? Will our dependence on AI erode our problem-solving skills and creativity?

AI as a Partner, Not a Replacement

The fear that AI will replace humans in various jobs is prevalent, yet this perspective often overlooks the true potential of artificial intelligence as a collaborative partner. Rather than viewing machines as competitors, we can see them as tools that augment our capabilities and enhance our performance. For example, in the realm of healthcare, AI can analyze patient data and suggest treatment options, but it is ultimately the doctor's compassion, experience, and judgment that guide patient care. This symbiotic relationship allows healthcare professionals to focus more on what they do best—caring for patients—while leveraging AI's strengths in data processing.

In creative industries, AI has been embraced as a collaborative partner, assisting artists, musicians, and writers in exploring new horizons. Tools that generate music, suggest plotlines, or create visual art based on prompts can serve as inspiration for human creators, pushing the boundaries of imagination. When humans and machines collaborate, the results can be extraordinary, blending the analytical prowess of AI with the emotional depth of human creativity. The partnership allows us to explore uncharted territories and redefine what's possible, ultimately leading to innovations that benefit society as a whole.

However, this collaboration requires a mindset shift. We must cultivate a culture of teamwork where humans and machines work together harmoniously. This involves training and education that emphasize not only technical skills but also interpersonal skills, creativity, and emotional intelligence—qualities that machines cannot replicate. By fostering this collaborative spirit, we can create a future where AI serves as an extension of human capabilities, enhancing our lives rather than threatening our livelihoods.

The Social Impact of Living with Machines

As AI technology permeates our lives, its influence on social interactions and relationships cannot be underestimated. While AI has the potential to enhance communication, streamline processes, and foster connections, it also poses challenges that warrant careful consideration. For instance, AI-driven platforms enable global connectivity, allowing people from diverse backgrounds to share ideas and collaborate on projects. However, this hyper-connectivity can sometimes lead to a sense of isolation, as individuals may rely more on screens than on face-to-face interactions.

Moreover, the algorithms that govern social media platforms often prioritize content that generates engagement, which can create echo chambers and reinforce existing biases. As users are exposed to curated content that aligns with their beliefs, meaningful dialogue can diminish, leading to polarization and a lack of understanding across different perspectives. The challenge lies in ensuring that AI fosters genuine connections and open discourse rather than stifling it.

The impact of AI on our social fabric extends beyond communication. In workplaces, AI can enhance productivity by automating repetitive tasks, allowing employees to focus on more creative and strategic endeavors. Yet, this shift can also create anxiety about job security and the need for continuous reskilling. Workers may feel the pressure to adapt quickly to new technologies, leading to stress and uncertainty in their professional lives.

To navigate these social dynamics, it is crucial for individuals, organizations, and communities to establish a framework for healthy interactions with AI. This includes promoting digital literacy, encouraging critical thinking, and fostering an inclusive environment where diverse voices are heard. By consciously shaping the role of AI in our social interactions, we can harness its benefits while mitigating its potential downsides, ultimately fostering a society where technology serves to enhance our humanity rather than diminish it.

The Psychological Effects of Living with Machines

The rise of AI and machine learning is also having profound psychological effects on individuals. As machines handle more of our tasks, there is the potential for humans to experience both positive and negative psychological impacts. On one hand, AI can reduce stress by handling mundane tasks, leaving people with more time for creative and fulfilling activities. On the other hand, the increasing automation of jobs and daily tasks can lead to anxiety about job security, a sense of loss of control, and even feelings of inadequacy.

Moreover, the integration of AI into personal lives has raised concerns about human dependence on machines for emotional support. AI-driven virtual companions or mental health chatbots are being developed to provide emotional assistance, but they cannot truly replace human empathy. As we navigate this new relationship with machines, we must be mindful of the potential emotional and psychological effects and ensure that AI tools are used to enhance, rather than diminish, our emotional well-being.

Ethical Considerations in Coexisting with Machines

As we integrate machines more deeply into our lives, ethical questions inevitably arise. Who is responsible when a machine makes a mistake? How do we ensure that AI technologies are deployed in ways that respect privacy, autonomy, and human dignity? These questions become even more critical as AI systems are given greater autonomy in decision-making, from self-driving cars to military drones.

Coexisting with machines requires us to establish clear ethical guidelines for their development and use. Machines are, ultimately, tools created by humans, and their impact on society depends on how we design and deploy them. It is up to us to ensure that AI is aligned with human values and that it serves the greater good.

Adapting to a World Shared with Machines

As AI continues to evolve, humans will need to adapt to a world where machines are constant companions. This adaptation will require both a shift in mindset and practical changes in how we live and work. Education systems will need to prepare future generations for a world where collaboration with AI is the norm. Governments will need to create regulatory frameworks that protect citizens' rights while fostering innovation. And individuals will need to develop new skills—such as digital literacy, emotional intelligence, and creative problem-solving—that will be essential in an AI-driven world.

Coexisting with machines also means rethinking what it means to be human. As machines take over more tasks, we may find ourselves free to focus on what makes us unique: our ability to create, to empathize, and to solve complex problems. Rather than seeing machines as competitors, we can view them as partners in building a future where humans and AI work together to address global challenges, from climate change to healthcare.

Conclusion: Embracing a Symbiotic Relationship

Coexisting with machines is not about choosing between humans and AI—it's about finding a way to work together in a way that benefits society as a whole. AI has the potential to enhance our lives, but it also presents challenges that must be carefully navigated. As we share our world with machines, we must ensure that this relationship is symbiotic, where humans remain at the center of decision-making, creativity, and ethical considerations.

The future will not be defined by machines alone but by how well we, as humans, can integrate them into our lives in a way that promotes progress while preserving our humanity. By approaching AI with a balanced perspective, we can coexist with machines in a world that is more efficient, innovative, and human-centered.

Chapter 14: The Future of Human Identity in an AI-Driven World

As artificial intelligence continues to evolve and reshape our lives, it invites profound questions about the very essence of human identity. Who are we in a world where machines can mimic our behaviors, preferences, and even emotions? This chapter explores how AI challenges traditional notions of identity, individuality, and what it means to be human in an increasingly digital landscape.

Redefining Identity in the Age of AI

Historically, human identity has been shaped by our experiences, relationships, and cultural contexts. However, as AI systems learn from vast datasets and interact with us on a personal level, the lines between human and machine begin to blur. With algorithms capable of analyzing our behaviors and predicting our preferences, AI can tailor experiences in ways that feel deeply personal. Yet, this personalization raises questions about authenticity. If machines know us better than we know ourselves, what happens to our sense of self?

One striking example is the rise of virtual avatars and digital personas. In virtual environments, individuals can craft identities that differ from their real-world selves, often exploring facets of their personalities that they may not express offline. While this can lead to empowerment and self-discovery, it can also create confusion about who we are when our online and offline identities do not align. As we increasingly engage with AI in social media, gaming, and virtual reality, the concept of identity becomes a complex interplay of our true selves and the personas we choose to project.

The Impact of AI on Self-Perception and Social Identity

The integration of AI into daily life also impacts our self-perception and how we relate to others. Social media platforms, driven by AI algorithms, often promote idealized representations of life, leading to comparison and competition. This can foster feelings of inadequacy and influence how individuals perceive their worth based on likes, shares, and follower counts. The pressure to conform to these curated standards can overshadow the authentic self, leading to an identity crisis for many.

Moreover, AI's role in shaping our social identity cannot be overlooked. As AI systems analyze and categorize individuals based on their online behavior, they create data-driven profiles that can influence how we are perceived by others. For instance, targeted advertising and content recommendations can reinforce stereotypes and biases, shaping public perception in ways that can be detrimental to marginalized groups. This raises ethical concerns about the potential for AI to perpetuate discrimination and limit the complexity of human identity to mere data points.

Chapter 14: The Future of Human Identity in an AI-Driven World

Cultural Implications of an AI-Infused Identity

The cultural implications of an AI-driven world are vast. As technology becomes a dominant force in shaping our identities, cultural narratives may shift. Traditional markers of identity—such as nationality, ethnicity, and religion—may evolve or lose significance as individuals increasingly identify with digital communities that transcend geographical boundaries. This shift can foster global solidarity and collaboration, as people connect over shared interests and values rather than cultural divisions.

However, this global connectivity also poses challenges to cultural heritage and individuality. As AI-generated content floods the digital landscape, there is a risk of homogenization, where unique cultural expressions are diluted in favor of mainstream trends. The challenge lies in balancing the benefits of global interconnectedness with the preservation of cultural diversity and individual expression.

The Search for Authenticity in a Digital Age

In the face of these challenges, the quest for authenticity becomes increasingly vital. As AI-driven tools and technologies become more prevalent, individuals must grapple with how to maintain their sense of self amid a sea of digital personas and curated realities. This search for authenticity may manifest in various ways, from conscious efforts to unplug and engage in genuine human interactions to seeking out experiences that prioritize realness over virtuality.

Furthermore, as we navigate this digital age, the importance of emotional intelligence, creativity, and critical thinking will become more pronounced. While AI can analyze data and automate tasks, these uniquely human traits will continue to be essential in fostering connections, understanding ourselves, and expressing our identities. By prioritizing these qualities, we can build a future where technology enhances our humanity rather than diminishes it.

Preparing for a Future of Fluid Identities

As we advance into a world increasingly shaped by AI, it's crucial to prepare for fluid identities. Education systems, workplaces, and communities must adapt to cultivate resilience, adaptability, and self-awareness. Encouraging individuals to reflect on their identities and values will help them navigate the complexities of an AI-driven world while maintaining a strong sense of self.

Additionally, fostering open dialogues about the ethical implications of AI on identity empowers individuals to take an active role in shaping their digital experiences. By advocating for transparency in AI algorithms and promoting inclusive technologies, we can create a more equitable digital landscape that respects and celebrates the diversity of human identities.

Chapter 14: The Future of Human Identity in an AI-Driven World

Conclusion: Embracing Our Humanity in a Technological Future

The future of human identity in an AI-driven world is a landscape filled with both possibilities and challenges. As we embrace the advancements of artificial intelligence, we must remain vigilant about preserving our humanity and authenticity. Our identities are not merely defined by algorithms or digital footprints; they are shaped by our experiences, emotions, and connections with others.

By recognizing the profound impact of AI on our sense of self, we can navigate this new terrain with intention and purpose. Embracing our humanity while coexisting with machines will empower us to define our identities on our own terms, fostering a future where technology enhances the richness of human experience rather than diminishes it.

PART FOUR:

THE ROAD AHEAD

Chapter 15: AI for Good: Potential and Possibility

As we continue to navigate the complexities of artificial intelligence, it is essential to explore the positive impact that AI can have on society. While concerns about its implications are valid, the potential of AI to address pressing global challenges and enhance the human experience cannot be overlooked. This chapter delves into the ways AI can be harnessed for good, highlighting its potential and the possibilities that lie ahead.

Addressing Global Challenges with AI

AI has emerged as a powerful tool in tackling some of the world's most significant challenges, from climate change and healthcare to education and poverty alleviation. For instance, in the realm of climate action, AI-driven models can analyze vast amounts of environmental data to predict weather patterns, assess natural disaster risks, and optimize resource management. By providing insights that inform policy decisions and conservation efforts, AI can play a crucial role in fostering a sustainable future.

In healthcare, AI has the potential to revolutionize diagnostics and treatment. Machine learning algorithms can analyze medical images, identify patterns, and assist in early disease detection, leading to improved patient outcomes. Additionally, AI can personalize treatment plans by considering individual patient data, ensuring that healthcare is more effective and accessible. This transformative potential extends to mental health, where AI-powered applications can offer support through chatbots, providing users with immediate access to resources and guidance.

Enhancing Education and Learning Experiences

Education is another area where AI can make a significant difference. Intelligent tutoring systems can adapt to individual learning styles and paces, providing personalized instruction that meets students where they are. This adaptability not only enhances learning outcomes but also fosters a love for learning by catering to each student's unique needs.

Moreover, AI can democratize education by making quality resources accessible to learners worldwide. With AI-driven platforms, students from diverse backgrounds can access educational materials, participate in virtual classrooms, and engage with expert educators, regardless of their geographical location. This shift has the potential to bridge educational gaps and empower individuals with the knowledge and skills needed to thrive in a rapidly changing world.

Fostering Community and Social Good

AI can also be leveraged to foster community engagement and social good. Nonprofit organizations are increasingly utilizing AI to identify areas of need and allocate resources more effectively. By analyzing data on social issues, AI can help organizations target their efforts where they will have the greatest impact, whether that be in disaster relief, food distribution, or healthcare access.

Furthermore, AI-driven platforms can facilitate connections within communities, empowering individuals to collaborate on local initiatives and support one another. For example, neighborhood apps can use AI to match volunteers with residents in need of assistance, fostering a sense of community and collective responsibility. In this way, AI can enhance social bonds and encourage civic engagement, driving positive change at the grassroots level.

Ethical Considerations and Responsible AI Development

While the potential of AI for good is vast, it is imperative to approach its development and implementation with ethical considerations at the forefront. As we harness AI's capabilities, we must ensure that it is used responsibly and equitably. This includes addressing biases in algorithms, safeguarding privacy, and prioritizing transparency in AI systems.

Collaborative efforts between governments, industries, and civil society are essential to create a framework for responsible AI development. By establishing guidelines and best practices, we can ensure that AI is deployed in ways that benefit all members of society, minimizing the risks associated with its use. This collaborative approach will foster trust and accountability, enabling us to leverage AI for the greater good.

A Vision for the Future

The future of AI holds immense potential for positive change. By embracing AI as a force for good, we can reimagine our world, creating solutions to pressing challenges while enhancing the quality of life for individuals and communities. This vision requires a commitment to innovation, collaboration, and ethical practices that prioritize human well-being.

As we look ahead, we must also remain vigilant and proactive in addressing the complexities that arise with AI integration. By fostering an inclusive dialogue about its impact and possibilities, we can shape a future where AI serves as a partner in our quest for progress and social good. Together, we can unlock the potential of artificial intelligence to create a better, more equitable world for all.

Conclusion: Harnessing AI for a Brighter Tomorrow

In conclusion, AI has the potential to transform our world for the better, addressing global challenges, enhancing education, and fostering community engagement. By focusing on responsible development and ethical considerations, we can harness AI as a powerful ally in our pursuit of social good. The possibilities are vast, and the road ahead is filled with hope, innovation, and the promise of a brighter tomorrow.

Chapter 16: Can We Survive the AI Revolution?

As we stand at the precipice of an AI revolution, the question looms large: can humanity not only survive but thrive in a world increasingly dominated by artificial intelligence? The rapid advancements in AI technology prompt both excitement and anxiety, as we grapple with the implications of these changes on our lives, jobs, and society as a whole. This chapter explores the challenges and opportunities presented by the AI revolution and contemplates what it means for our collective future.

Understanding the AI Revolution

The AI revolution refers to the significant transformation in how we interact with technology, characterized by the integration of AI into various aspects of life and work. From automated processes in industries to AI-driven personal assistants in our homes, these advancements are reshaping the fabric of our daily experiences. While the potential benefits are substantial, they are accompanied by challenges that necessitate careful consideration.

One of the most pressing concerns is the displacement of jobs. Many sectors, including manufacturing, retail, and even professional services, are witnessing automation that threatens traditional employment. While AI has the potential to create new jobs, the transition may not be smooth for those whose roles are rendered obsolete. This disruption raises critical questions about our economic structures and the future of work. How do we prepare for a workforce that must adapt to the rapid pace of technological change?

The Importance of Resilience and Adaptability

In the face of the AI revolution, resilience and adaptability emerge as vital traits for individuals and societies. Embracing lifelong learning and skills development will be essential for navigating an evolving job landscape. Education systems must pivot to focus on critical thinking, creativity, and emotional intelligence—skills that AI cannot replicate. By fostering these competencies, we can empower individuals to thrive alongside AI rather than be sidelined by it.

Moreover, societal resilience will depend on our ability to adapt to the changes brought about by AI. Policymakers, businesses, and communities must work together to create frameworks that support workers during transitions, ensuring access to retraining and upskilling opportunities. Social safety nets may need to evolve to accommodate shifts in employment patterns, safeguarding those affected by automation while encouraging innovation and entrepreneurship.

Ethical Considerations in the AI Landscape

Surviving the AI revolution also hinges on ethical considerations in the development and deployment of AI technologies. As AI systems become more pervasive, ensuring fairness, accountability, and transparency is paramount. Bias in algorithms can perpetuate discrimination, while lack of accountability can lead to misuse or harm. By prioritizing ethical practices, we can cultivate trust in AI and mitigate potential risks associated with its adoption.

Public engagement is crucial in shaping the ethical landscape of AI. Encouraging diverse voices in discussions about AI governance can lead to more inclusive decision-making processes. By involving communities, advocacy groups, and experts from various fields, we can create guidelines that reflect societal values and protect individual rights. This collaborative approach will be instrumental in navigating the complex moral terrain of the AI revolution.

The Potential for a Collaborative Future

While the challenges of the AI revolution are significant, the potential for a collaborative future also shines through. AI can enhance human capabilities, augmenting our strengths rather than replacing them. By viewing AI as a partner rather than a competitor, we can harness its capabilities to solve complex problems, drive innovation, and improve quality of life.

For instance, in healthcare, AI can assist doctors in diagnosing diseases more accurately, enabling more personalized treatment plans. In education, AI can support teachers by providing tailored resources for students, fostering more effective learning environments. These collaborative opportunities highlight the symbiotic relationship between humans and machines, where each can complement the other's strengths.

Navigating the Road Ahead

To survive the AI revolution, we must embrace change while remaining vigilant about its implications. A proactive approach involves anticipating challenges and preparing solutions before they arise. This means fostering a culture of innovation, encouraging interdisciplinary collaboration, and investing in research that prioritizes human-centered design.

Additionally, as citizens of an increasingly interconnected world, we must advocate for policies that promote equitable access to AI technologies. Ensuring that the benefits of AI are shared broadly will be crucial in preventing exacerbation of existing inequalities. By prioritizing inclusivity and accessibility, we can work toward a future where AI serves as a tool for collective progress rather than division.

Conclusion: A Call to Action for the Future

In conclusion, the question of whether we can survive the AI revolution hinges on our collective choices and actions. By cultivating resilience, fostering ethical practices, and embracing collaboration, we can navigate the complexities of an AI-driven world. The future is not predetermined; it is shaped by our responses to the challenges and opportunities before us.

As we embark on this journey, let us commit to being active participants in shaping the trajectory of the AI revolution, ensuring that it serves humanity's best interests. Together, we can build a future that celebrates the synergy of human ingenuity and artificial intelligence, paving the way for a society that thrives in the face of change.

Chapter 17: Your Role in the AI Future

As artificial intelligence continues to shape our world, it is essential to recognize that each individual has a unique role in navigating this transformative landscape. The choices we make, the skills we develop, and the values we uphold will all contribute to shaping an AI future that aligns with our collective aspirations. This chapter explores how you can actively engage in the evolving dialogue around AI and become a responsible participant in this revolution.

Embracing Lifelong Learning

The first step in preparing for an AI-driven future is embracing lifelong learning. As the pace of technological advancement accelerates, the demand for new skills and knowledge will grow. It's crucial to stay informed about emerging trends in AI and related fields. Whether through formal education, online courses, or self-directed study, investing in your education will empower you to adapt to the changes ahead.

Focus on developing both technical skills and soft skills. While understanding AI technologies—such as machine learning, data analysis, and programming—can enhance your employability, soft skills like critical thinking, creativity, and emotional intelligence are equally vital. These attributes enable you to navigate complex situations, collaborate effectively with others, and innovate in ways that machines cannot replicate.

Cultivating a Responsible Digital Citizenship

As AI systems become integrated into our daily lives, responsible digital citizenship becomes paramount. This means being aware of how your online actions can impact others and engaging with technology in a manner that promotes ethical use. Be mindful of the information you share, the sources you trust, and the content you consume. In a world where misinformation can spread rapidly, critical evaluation of information is essential.

Moreover, consider the ethical implications of AI technologies. As a consumer, you can advocate for transparency and accountability in the products and services you use. Support companies that prioritize ethical AI practices, and participate in discussions about the societal impacts of AI. By being an informed citizen, you can contribute to shaping policies that prioritize human rights and ethical standards in AI development.

Participating in the AI Dialogue

Your voice matters in the ongoing dialogue surrounding AI. Engage with your community, share your thoughts, and participate in discussions about the future of AI. This could involve attending local meetups, joining online forums, or contributing to blogs and social media conversations. By sharing your perspective, you can help shape the narrative around AI and encourage others to consider the broader implications of these technologies.

Additionally, seek opportunities to collaborate with others who share your interests in AI. Whether you join a local tech group, participate in hackathons, or contribute to open-source projects, collaborative efforts can lead to innovative solutions and a deeper understanding of AI's potential. By working together, you can amplify your impact and contribute to a more inclusive and equitable AI landscape.

Fostering Inclusivity and Diversity

In the pursuit of an AI future that benefits everyone, it is crucial to advocate for inclusivity and diversity. The development of AI technologies should reflect a wide range of perspectives, experiences, and backgrounds. Encourage diverse voices in discussions about AI, whether in the workplace, educational settings, or community forums. This inclusivity fosters creativity and innovation, ensuring that AI solutions address the needs of various demographics.

Moreover, consider how AI can be leveraged to promote social good. Explore initiatives that use AI to tackle pressing societal challenges—such as climate change, healthcare access, and education disparities. By engaging with projects that prioritize the common good, you can play a role in creating a future where AI serves as a tool for positive change.

Becoming an Advocate for Ethical AI

As AI continues to advance, the importance of ethical considerations cannot be overstated. As an individual, you can advocate for responsible AI practices by supporting policies that promote fairness, accountability, and transparency in AI development. This advocacy can take various forms, from signing petitions and engaging with lawmakers to participating in public forums that address AI ethics.

Educate yourself and others about the ethical dilemmas surrounding AI. Encourage conversations about issues such as algorithmic bias, data privacy, and the implications of automation on employment. By raising awareness and fostering dialogue, you contribute to a collective understanding of the ethical landscape surrounding AI technologies.

Conclusion: Your Future, Your Responsibility

In conclusion, your role in the AI future is not passive; it is an active, dynamic engagement that shapes the trajectory of technological advancement. By embracing lifelong learning, cultivating responsible digital citizenship, participating in the AI dialogue, fostering inclusivity, and advocating for ethical AI, you contribute to a future that aligns with our shared values.

As we move forward into an AI-driven world, remember that the choices you make today will impact the society of tomorrow. By taking responsibility for your role in this revolution, you can help ensure that AI serves as a force for good, enhancing human potential and contributing to a more equitable and sustainable future for all.

Epilogue: The Next Chapter for Humanity

As we conclude this exploration of artificial intelligence and its implications for humanity, it's essential to reflect on what lies ahead. The rapid evolution of AI is not merely a technological advancement; it represents a profound shift in the way we live, work, and interact with one another. In this epilogue, we will examine the potential futures that await us, the choices we must make, and the responsibility we bear in shaping a world where technology enhances the human experience.

A Tapestry of Possibilities

The future is a vast tapestry woven from countless threads of possibility. AI holds the promise of revolutionizing industries, transforming healthcare, enhancing education, and addressing pressing global challenges such as climate change and resource scarcity. However, these possibilities are not guaranteed; they depend on the choices we make as individuals, communities, and societies.

Imagine a world where AI assists us in making informed decisions, where smart technologies help us conserve energy, manage our health, and connect with one another in meaningful ways. Picture classrooms where AI personalizes learning experiences, allowing every student to thrive at their own pace. Envision a society where AI-driven solutions address societal inequalities, providing access to education, healthcare, and economic opportunities for all.

Yet, this vision is only achievable if we collectively commit to guiding the development and deployment of AI in ways that prioritize human well-being and ethical considerations. The path forward requires collaboration, creativity, and a willingness to embrace change while remaining vigilant about its implications.

The Human Element in a Technological Age

Amidst the advancements in AI, the human element must remain at the forefront. Technology should not replace human connection; rather, it should enhance our ability to empathize, collaborate, and innovate. As we integrate AI into our lives, we must prioritize the values that define us as humans—compassion, creativity, and the pursuit of knowledge.

In this era of rapid change, we must also recognize the importance of mental and emotional well-being. As technology takes on more roles in our daily lives, it is essential to foster a sense of purpose, community, and belonging. Encouraging meaningful relationships, nurturing creativity, and cultivating mindfulness will help us navigate the challenges that lie ahead.

Navigating Ethical Waters

As AI continues to evolve, we are faced with ethical dilemmas that require thoughtful consideration. Issues such as data privacy, algorithmic bias, and the impact of automation on employment demand our attention. The decisions made today regarding the ethical use of AI will have long-lasting implications for future generations.

To navigate these waters, we must advocate for transparent practices and inclusive decision-making processes. Engaging diverse voices in discussions about AI ethics will lead to more equitable outcomes and solutions that reflect the values of our society. It is our responsibility to ensure that AI serves the greater good and contributes to a future that upholds human dignity and rights.

A Call to Action for the Next Generation

As we look to the future, the responsibility for shaping the next chapter for humanity lies not only with policymakers and industry leaders but with each one of us. The choices we make today will impact the world we leave for future generations. Embracing lifelong learning, fostering inclusivity, and advocating for ethical practices are all critical actions we can take to contribute to a positive AI future.

For the younger generations, the opportunities are boundless. You are the innovators, thinkers, and leaders of tomorrow. Embrace the potential of AI, explore its applications, and harness its capabilities to create solutions for the challenges we face. Your unique perspectives and fresh ideas will play a pivotal role in shaping a future where technology and humanity coexist harmoniously.

Conclusion: Together Towards a Brighter Future

In conclusion, the next chapter for humanity is unwritten, and we are the authors of this narrative. By embracing the possibilities of AI while remaining grounded in our values, we can shape a future that is not only technologically advanced but also compassionate, inclusive, and just.

As we embark on this journey into the unknown, let us remember that the essence of humanity lies in our ability to connect, empathize, and uplift one another. Together, we can navigate the

AI & You: Will Humanity Survive the Future We're Building?

complexities of an AI-driven world, ensuring that technology enhances our lives and empowers us to build a better future for all.

The story of humanity's relationship with AI is just beginning, and it is up to us to write the next chapter—one that reflects our highest aspirations and our deepest commitments to one another.

Feel free to let me know if you want any changes or additional elements in this chapter!

About the Author

Faisal Siddiqi is a distinguished corporate leader, keen observer, avid reader, and philanthropist, renowned for his strategic vision and impactful leadership. With a remarkable track record in crafting and executing business strategies, Faisal has driven organizational success across multiple sectors, particularly in telecommunications. His expertise in global team-building, project management, and stakeholder relations has resulted in significant returns and operational improvements throughout his career. His ability to lead cross-functional teams and forge strong partnerships has been a hallmark of his professional journey.

Beyond his corporate accomplishments, Faisal is an avid analyst and observer of Technology, global politics and current affairs, constantly seeking to understand and interpret the evolving dynamics that shape our world. His deep understanding of these issues allows him to offer unique insights into the intersection of business, politics, and society.

Faisal's commitment to social good is equally noteworthy. As a philanthropist, he dedicates much of his time and resources to initiatives aimed at improving education, healthcare, and the overall welfare of underserved communities. His belief in giving back to society informs much of his work, both within and outside the business realm.

A passionate reader with an insatiable curiosity, Faisal combines his love of literature with his professional experience, often blending knowledge from a variety of fields to provide thoughtful commentary and analysis. His writing reflects his multifaceted approach to leadership, his dedication to societal advancement, and his belief in the power of informed action to effect positive change.